T0262024

Encyclopedia of Remote Sensing: Sensors and Platforms

Volume II

Encyclopedia of Remote Sensing: Sensors and Platforms
Volume II

Edited by **Matt Weilberg**

New York

Published by Callisto Reference,
106 Park Avenue, Suite 200,
New York, NY 10016, USA
www.callistoreference.com

Encyclopedia of Remote Sensing: Sensors and Platforms
Volume II
Edited by Matt Weilberg

International Standard Book Number: 978-1-63239-289-3 (Hardback)

Printed in the United States of America.

Contents

Preface

The world is advancing at a fast pace like never before. Therefore, the need is to keep up with the latest developments. This book was an idea that came to fruition when the specialists in the area realized the need to coordinate together and document essential themes in the subject. That's when I was requested to be the editor. Editing this book has been an honour as it brings together diverse authors researching on different streams of the field. The book collates essential materials contributed by veterans in the area which can be utilized by students and researchers alike.

This book gives a comprehensive account of various areas related to remote sensing. It attempts to cover major aspects of remote sensing techniques and platforms by providing detailed studies about sensors and platforms and new techniques required for data processing. It includes contributions of renowned researchers, experts and practitioners in this field from all over the world. Their collective expertise impart significant knowledge and serve as a valuable reference for researchers, students and other interested individuals in this field.

Each chapter is a sole-standing publication that reflects each author's interpretation. Thus, the book displays a multi-facetted picture of our current understanding of application, resources and aspects of the field. I would like to thank the contributors of this book and my family for their endless support.

Editor

Sensors and Platforms

COMS, the New Eyes in the Sky for Geostationary Remote Sensing

Han-Dol Kim et al.[*]
Korea Aerospace Research Institute (KARI)
Republic of Korea

1. Introduction

With its successful launch on June 26, 2010, the Communication, Ocean, and Meteorological Satellite (COMS) is currently in the early stage of normal operation for the service to the end users, exhibiting exciting and fruitful performances including the image data from the two on-board optical sensors, Meteorological Imager (MI) and Geostationary Ocean Color Imager (GOCI), and the experimental Ka-band telecommunication. This chapter gives a comprehensive overview of COMS in terms of its key design characteristics, current status of in-orbit performances and its implied role in the geostationary remote sensing, and discusses its potential application and contribution to the world remote sensing community.

2. COMS: Description and overview

COMS is a multi-purpose, multi-mission, geostationary satellite. It has been designed and developed by the joint effort of EADS Astrium and Korea Aerospace Research Institute (KARI), and launched by Ariane 5 ECA L552 V195 of Arianespace on 21:41 (UTC) of June 26 2010. COMS is the first South Korean multi-mission geostationary satellite, and also the first 3-axis stabilized geostationary satellite ever built in Europe for optical remote sensing.

The In Orbit Testing (IOT) of COMS was completed early part of 2011, and since then the satellite has been being successfully operated by KARI for the benefits of all 3 end users: the Korean Meteorological Administration (KMA), the Korea Ocean Research & Development Institute (KORDI) and the Electronics & Telecommunications Research Institute (ETRI).

2.1 COMS overview

COMS is a single geostationary satellite fulfilling 3 rather conflicting missions as follows:

- A meteorological mission by MI

[*] Gm-Sil Kang[1], Do-Kyung Lee[1], Kyoung-Wook Jin[1], Seok-Bae Seo[1], Hyun-Jong Oh[2], Joo-Hyung Ryu[3], Herve Lambert[4], Ivan Laine[4], Philippe Meyer[4], Pierre Coste[4] And Jean-Louis Duquesne[4]
[1]*Korea Aerospace Research Institute (KARI), Republic of Korea*
[2]*Korea Meteorological Administration (KMA), Republic of Korea*
[3]*Korea Ocean Research & Development Institute (KORDI), Republic of Korea*
[4]*EADS Astrium, France*

- An ocean imager mission by GOCI
- An experimental Ka band telecommunication mission

MI is the common imager with the flight heritage from the later series of GOES and MTSAT satellites, and GOCI is the world' 1st ocean color imager to be operated in the geostationary orbit which has been newly developed for the COMS mission. The spacecraft launch mass is 2460 kg and the size is 2.6 m x 1.8 m x 2.8 m in stowed configuration. The orbital location is 128.2°E, mission lifetime is 7.7 years and design lifetime is 10 years.

Fig. 1 shows COMS both in stowed and deployed configurations, where the MI and GOCI optical instruments located on the earth looking satellite floor can be found with both MODCS (Meteorology and Ocean Data Communication System) antenna and the two small telecommunication Ka band reflectors, along with the COMS flight model during AIT.

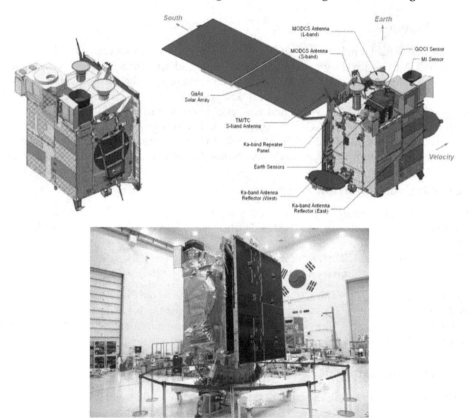

Fig. 1. COMS, in stowed and deployed configurations and the flight model during the final stage of AIT (Assembly, Integration and Test) at KARI

The following subsections give a succinct description of COMS system, in terms of its key design characteristics and its unique and salient features on the platform, with a little touch on its development history and with a certain emphasized details on GOCI, along with a brief description on the ground segment.

2.2 Description of COMS system

The COMS system consists of the space segment, which is made up of a COMS spacecraft bus with the three payloads, and the various systems of the ground segment, as depicted in the Fig. 2.

Images captured by MI and GOCI are first interleaved on board and downloaded in L band. Data are separated on ground; MI data are processed (radiometrically calibrated and geometrically corrected) and uploaded again in S-band to the satellite in two formats, LRIT (Low Rate Information Transmission) and HRIT (High Rate Information Transmission). These two new streams of data are again interleaved with the raw data and downloaded in L-band to end users by the satellite which acts as a specific data relay.

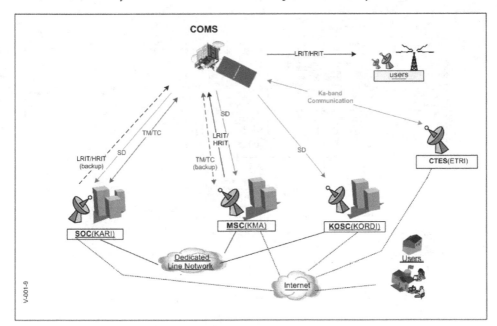

Fig. 2. COMS system overview

2.2.1 COMS spacecraft bus

The COMS spacecraft bus is based on EADS Astrium's Eurostar-3000 bus design. The satellite features a box-shaped structure, built around the two bi-propellant tanks. Imaging instruments and MODCS antennae are located on the Earth floor (Fig. 1). A single-winged solar array with 10.6 m² of GaAs cells is implemented on the south side, so as to keep the north wall in full view of cold space for the MI radiant cooler. The deployable Ka-band antenna reflectors are accommodated on the east and west walls.

The COMS spacecraft is 3-axis stabilized. Attitude sensing in normal mode is based on a hybridized Earth sensors (IRESs; Infra-Red Earth Sensors) and gyros (FOGs; Fiber Optic Gyros) concept; in addition, sun sensors are being used during 3-axis transfer operations. 5 reaction wheels (RDRs) and 7 thrusters (10 N) serve as actuators. Thrusters are also used for wheel off-loading and for orbit control. The apogee firing boosts are provided by a 440 N liquid apogee engine.

The key feature of COMS AOCS (Attitude and Orbit Control Subsystem) is the addition of EADS Astrium's newly developed FOGs, Astrix 120 HR. The FOG allowed the requested performance boost in terms of pointing knowledge and stability to already excellent Eurostar-3000 AOCS design and its performances.

The EPS (Electric Power Subsystem) makes use of GaAs solar cells and Li-ion batteries. A regulated power bus (50 V) distributes power to the various onboard applications through the power shunt regulator. During orbital eclipses, energy is provided by a 154 Ah Li-ion battery. The power at EOL (End Of Life) shall be greater than 2.5 KW.

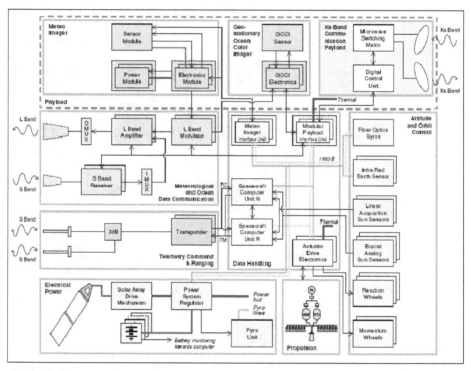

Fig. 3. Block diagram of COMS spacecraft functional architecture

The heart of the avionics architecture is implemented in hot redundant spacecraft computer units, based on 1750 standard processors with Ada object-oriented real-time software. A redundant MIL-STD-1553-B data bus serves as the main data path between the onboard units. Interface units are being used for the serial links, namely the actuator drive electronics with the bus units (including thermal control), the modular payload interface unit with the Ka-band communication payload, and the MI interface unit with the MI instrument.

A specific module (MODCS; Meteorology and Ocean Data Communication System) was developed for handling MI and GOCI images. It collects and transmits raw MI and GOCI data in L-band. HRIT/LRIT (High- and Low-Rate Information Transmission) formats are generated on the ground from the MI raw data, and uploaded to the satellite in S-Band and relayed in L-band to MI end users.

S-band is also used for satellite Telemetry and Telecommands.

2.2.2 MI

MI is a two-axis scan imaging radiometer from ITT. It senses basically the radiant and solar reflected energies from the Earth simultaneously and provides imagery and radiometric information of the Earth's surface and cloud cover. It features 1 visible (VIS) channel and 4 infra-red (IR) channels as a scanning radiometer. The design of it is derived from the GOES imager for COMS program.

No.	Channel	Wavelength(µm)	IFOV(µrad)	GSD(Km)	Dynamic Range
1	VIS	0.55~0.80	28	1	0~115% albedo
2	SWIR	3.50~4.00	112	4	110K~350K
3	WV	6.50~7.00	112	4	110K~330K
4	WIN1	10.3~11.3	112	4	110K~330K
5	WIN2	11.5~12.5	112	4	110K~330K

Table 1. Spectral channel characteristics of MI as requirement

MI consists of three modules; sensor module, electronics module, and power supply module. The sensor module contains a scan assembly, a telescope and detectors, and is mounted on spacecraft with the shields, louver and cooler for thermal control. The electronics module which has some redundant circuits performs command, control, signal processing and telemetry conditioning function. The power supply module contains power converters, fuses and power control for interfacing with the spacecraft power system with redundancy.

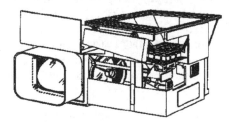 

Fig. 4. COMS MI sensor module, in design and flight model configurations

The servo-driven, two-axis gimbaled scan mirror of the MI reflects scene energy reflected and emitted from the Earth into the telescope of the MI as shown in the Fig. 5. The mirror scans the Earth with a bi-directional raster scan, which sweeps an 8 km swath along East-West (EW) direction and steps every 8 km along North-South (NS) direction. The area of the observed scene depends on the 2-dimensional angular range of the scan mirror movement. The scene radiance, collected by the scan mirror and the telescope, is separated into each spectral channel by dichroic beam splitters, which allow the geometrically-corresponding detectors of each channel to look at the same position on the Earth. Each detector converts

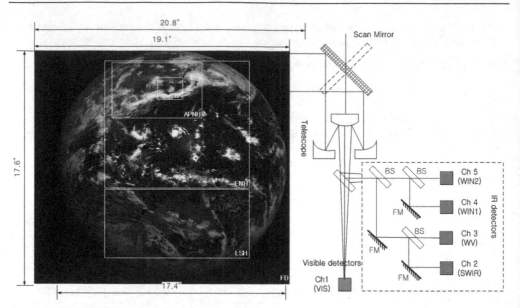

Fig. 5. MI Scan Frame and Schematic design of Optics (BS:Beam Splitter, FM:Folding Mirror, FD: Full Disk, APNH:Asia and Pacific in Northern Hemisphere, ENH:Extended Northern Hemisphere, LSH:Limited Southern Hemisphere, LA: Local Area)

the scene radiance into an electrical signal. The five channel detectors of the MI are divided into two sides, which are electrically redundant each other. Only one side operates at one time by choosing side 1 or side 2 electronics. The visible silicon detector array contains eight detector elements which are active simultaneously in the either side mode. Each visible detector element produces the instantaneous field of view (IFOV) of 28 μrad on a side, which corresponds to 1km on the surface of the Earth at the spacecraft's suborbital point. Each IR channel has two detector elements which are active simultaneously in the either side mode. The SWIR channel employs InSb detectors and the other IR channels use HgCdTe detectors. Each IR detector element produces the IFOV of 112 μrad on a side, which corresponds to 4km on the surface of the Earth at the spacecraft's suborbital point. The 8 visible detector elements and 2 IR detector elements produce the swath width (8 km) of one EW scan line respectively.

The passive radiant cooler with thermostatically controlled heater maintains the infrared detectors at one of the three, command-selectable, cryogenic temperatures. Visible light detectors are at the instrument ambient temperature. Preamplifiers convert low level outputs of all detectors into higher level, low impedance signals as the inputs to the electronics module. MI carries an on-board blackbody target inside of the sensor module for the in-orbit radiometric calibration of the IR channels. The blackbody target is located at the opposite direction to the nadir, so that the scan mirror is rotated 180 degrees in the NS direction from the imaging mode for the blackbody calibration. The full aperture blackbody calibration can be performed by the scan mirror's pointing at the on-board blackbody target via ground command or automatically. The albedo monitor is mounted in the sensor module to measure the in-orbit response change of the visible channel over the mission life.

It uses sunlight through a small aperture as a source. In addition to the radiometric calibration, an electrical calibration is provided to check the stability and the linearity of the output data of the MI signal processing electronics by using an internal reference signal. MI has the star sensing capability in the visible channel, which can be used for image navigation and registration purposes.

MI has three observation modes: global, regional and local modes, which are specialized for the meteorological missions. The global mode is for taking images of the Full Disk (FD) of the Earth. The regional observation mode is for taking images of the Asia and Pacific in North Hemisphere (APNH), the Extended North Hemisphere (ENH), and Limited Southern Hemisphere (LSH). The image of Limited Full Disk (LFD) area can be obtained by the combination of the images of ENH and LSH. The local observation mode is activated for Local Area (LA) coverage in the FD. The user interest of the MI observation areas for FD, APNH, ENH, LSH, LFD, and LA is shown in the Fig. 5.

2.2.3 GOCI

Geostationary Ocean Color Imager (GOCI), the first Ocean Colour Imager to operate from geostationary orbit, is designed to provide multi-spectral data to detect, monitor, quantify, and predict short term changes of coastal ocean environment for marine science research and application purpose. GOCI has been developed to provide a monitoring of Ocean Color around the Korean Peninsula from geostationary platforms in a joint effort by Korea Aerospace Research Institute (KARI) and EADS Astrium under the contract of Communication, Ocean, and Meteorological Satellite (COMS) of Korea.

2.2.3.1 GOCI mission overview

Main mission requirement for GOCI is to provide a multi-spectral ocean image of area around South Korea eight times per day as shown in Fig. 6. The imaging coverage area is 2500x2500 km² and the ground pixel size is 500x500 m² at centre of field, defined at (130°E -

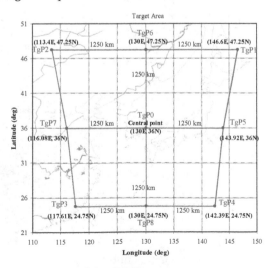

Fig. 6. Target observation coverage of the GOCI

36°N). Such resolution is equivalent to a Ground Sampling Distance (GSD) of 360 m in NADIR direction, on the equator. The GSD is varied over the target area because of the imaging geometry including the projection on Earth and the orbital position of the satellite. The GOCI spectral bands have been selected for their adequacy to the ocean color observation, as shown in Table 2.

Band	Center	Band-width	Main Purpose and Expected Usage
1	412 nm	20 nm	Yellow substance and turbidity extraction
2	443 nm	20 nm	Chlorophyll absorption maximum
3	490 nm	20 nm	Chlorophyll and other pigments
4	555 nm	20 nm	Turbidity, suspended sediment
5	660 nm	20 nm	Fluorescence signal, chlorophyll, suspended sediment
6	680 nm	10 nm	Atmospheric correction and fluorescence signal
7	745 nm	20 nm	Atmospheric correction and baseline of fluorescence signal
8	865 nm	40 nm	Aerosol optical thickness, vegetation, water vapour reference over the ocean

Table 2. GOCI spectral bands

2.2.3.2 GOCI design overview

The GOCI consists of a Main Unit and an Electronic Unit. Total GOCI Mass is below 78 Kg. Power needed is about 40W for the electronics plus about 60W for Main Unit thermal control. A Payload Interface Plate (PIP) is part of the Main Unit. It supports a highly stable full SiC telescope, mechanisms and proximity electronics. Fig. 7 shows the main unit which is integrated on the Earth panel of satellite through the PIP. The PIP is larger than the instrument to carry the satellite Infra-Red Earth Sensor (IRES).

The main unit includes an optical module, a two-dimensional Focal Plane Array (FPA) and a Front End Electronics (FEE). The optical module of GOCI consists of a pointing mirror, a Three Mirror Anastigmat (TMA) mirrors, a folding mirror, and a filter wheel. The FEE is attached near the FPA in order to amplify the detector signal with low noise before digitization.

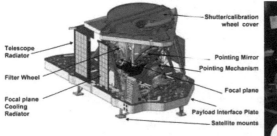

Fig. 7. Design configuration of GOCI main unit and its flight model configuration during integration phase (without MLI)

The shutter wheel is located in front of pointing mirror carrying four elements: shutter which will protect optical cavity during non-imaging period, open part for the ocean observation, Solar Diffuser (SD) and Diffuser Aging Monitoring Device (DAMD) for solar calibration. A Quasi Volumic Diffuser (QVD) has been chosen for the SD and the DAMD among several candidates because it is known to be insensitive to radiation environment. The on-board calibration devices prepared for integration are shown in Fig. 8. The SD covering the full aperture of GOCI is used to perform in-orbit solar calibration on a daily basis. Degradation of the SD over mission life is detected by the DAMD covering the partial aperture of GOCI.

| SD | DAMD | POM (without mirror) |

Fig. 8. On-board calibration devices SD, DAMD and pointing mirror mechanism POM

The pointing mirror is equipped with a 2-axis circular mechanism for scanning over observation area. Fig. 8 shows the GOCI pointing mechanism (POM). The pointing mirror is controlled to achieve a Line of Sight (LOS) corresponding to a center of a predefined slot on the Earth. The principle of the pointing mechanism is an assembly of two rotating actuators mounted together with a cant angle of about 1°, the top actuator carrying also the Pointing Mirror (PM) with the same cant angle. When rotating the lower actuator the LOS is moved on a circle and by rotating the second actuator, a second circle is drawn from the first one. It is thus possible to reach any LOS position inside the target area by choosing appropriate angle position on each circle. The mechanism pointing law provides the relation between rotation of both actuators and the LOS with a very high stability. This high accuracy pointing assembly used to select slots centers is able to position the instrument LOS anywhere within a 4° cone, with a pointing accuracy better than 0.03° (500 μrad). Position knowledge is better than 10 μrad (order of pixel size) thanks to the use of optical encoders. An incident light on the GOCI aperture is reflected by the pointing mirror and collected through the TMA telescope. Then the collected light goes to an optical filter through a folding mirror.

The eight spectral channels are obtained by means of a filter wheel which includes dark plate in order to measure system offset. Fig. 9 shows the filter wheel integrated with eight spectral filters without a protective cover. The FPA for GOCI, which is shown in Fig. 9, is a custom designed CMOS image sensor featuring rectangular pixel size to compensate for the Earth projection over Korea, and electron-optical characteristics matched to the specified instrument operations. The CMOS FPA having 1432 × 1415 pixels is passively cooled and regulated around 10°C. It is split into two modules which are electrically independent. The GOCI electronics unit, which is shown in Fig. 9, is deported on satellite wall about 1.5m from the GOCI main unit. It provides control of mechanisms (pointing mirror, shutter wheel, filter wheel), video data acquisition, digitization, mass memory and power.

Filter wheel   CMOS detector   Instrument Electronics Unit

Fig. 9. GOCI filter wheel without cover, CMOS detector package with temporary window and Electronics Unit

The imaging in GOCI is done in the step and stare fashion, passing along the 16 slots, as shown in the Fig. 10.

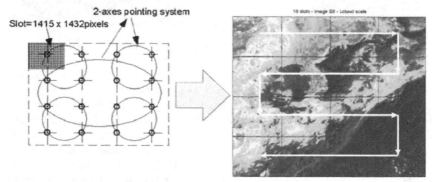

Fig. 10. GOCI imaging principle

2.2.4 COMS INR system

2.2.4.1 Overview of COMS INR system

Achieving and maintaining a good geo-localization of the images on the ground is an essential part of the geostationary remote sensing satellite for the untilization of the remote sensing data to be a meaningful and fruitful one. To this purpose, the Image Navigation and Registration (INR) system should be in place, and in COMS, a novel approach to INR was developed, allowing a-posteriori location of the images on the geoid based on automatic identification of landmarks and comparison with a reference database of specific terrestrial features such as small islands, capes, and lakes.

In this novel approach, INR is not directly dependent on the satellite and payload models and hence can avoid any indispensible modeling and prediction error in the process. The high reliance on the landmarks and the acquisition of sufficient number of good-quality landmarks, however, become the key part of the design in this approach and such acquisition must be secured for this approach to be practically successful. In COMS INR, excellent landmark matching algorithm, fine-tuning of configuration parameters during IOT and the fine-tuning of newly established landmark database with ample landmark sites at the final phase of IOT rendered such acquisition of sufficient number of good landmarks.

Fig. 11 shows the overall architecture of COMS INR. All the processing are done on ground except for the long term image motion compensation (LTIMC) and as can be seen here, the whole INR system is operated in close conjunction with the AOCS.

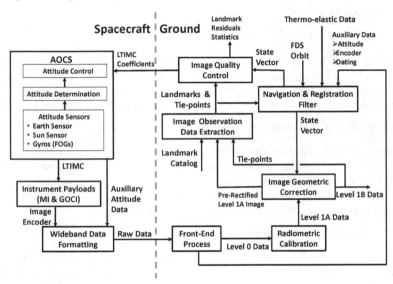

Fig. 11. COMS INR overall architecture

2.2.4.2 Description of COMS INR system and processing

In this section, the description of each module and each processing which comprises the whole COMS INR system, as shown in the Fig. 10, is provided.

2.2.4.2.1 Space Segment INR

Attitude determination

The on-board attitude determination estimates the spacecraft attitude from attitude sensors measurements (IRESs, Sun Sensors and FOGs) through filtering process. This process is performed at 10 Hz and sub-sampled at 1 Hz to insert into the MI wideband telemetry for use by the Navigation and Registration Filter Module on ground.

Attitude control

The on-board attitude control loop actuates momentum wheels, solar array, and thrusters. The control loop is designed to be robust to effect of disturbances on MI & GOCI field of views. Disturbances include:

- Diurnal attitude pointing perturbation due to thermo-elastic distortion and solar torques.
- Thruster firing for station keeping and wheel off-loading.

Long term image motion compensation (LTIMC)

The on-board LTIMC is used to compensate pointing bias and long term evolutions (seasonal, ageing) to keep the area to be observed within the MI & GOCI field of views.

Wideband data formatting

Wideband data consists of MI & GOCI imagery/telemetry and AOCS auxiliary attitude data.

2.2.4.2.2 Ground Segment INR

The image observation data extraction module

This module gathers all functions of data extraction from images: cloud cover detection, landmark detection from image/database matching, multi-temporal tie point detection from image/image matching, and multi-spectral tie point detection from band-to-band matching. A first "pre - rectification" of Level 1A images allows retrieving 2D local image coherence.

The navigation and registration filter module

This module gathers all functions of geometric models: localization model (including focal plane and scan mirror models), navigation filter, landmarks or tie points position prediction. This module performs state vector estimation through a hybridization filter that combines landmarks, thermo-elastic, orbit, and gyros data in the way that minimizes criteria on landmarks (for navigation) or tie points (for registration) residuals.

The image geometric correction module

This module gathers all functions relative to image resampling and Modulation Transfer Function (MTF) compensation. For each pixel of an image, the state vector allows computing the shift between raw geometry and reference geometry. Each pixel of the Level 1B image is computed through radiometric interpolation with respect to the neighbouring pixels around its corresponding pixel in the Level 1A image.

The image quality control module

Once the state filter estimation is performed, ground pixels corresponding to landmarks are localized. The result difference with respect to the landmark known position is called "residual". It can be done on the landmark used for navigation, but also on "reference landmarks" which are used for the navigation accuracy control filter. All computed residuals are stored for further statistics. The statistics (average, standard deviation, max value) on residuals within the image gives instantaneous INR performance. The statistics over a set of image during a certain period gives INR performance relative to the period. The statistics relative to a specific landmark over a certain period gives information on quality and reliability for that landmark. This result will be used to periodically update landmark database with confidence rate that has to be taken into account for better accuracy of the navigation filter. All statistics are also computed with respects to context: date, time, cloud conditions.

2.2.5 COMS ground segment

The COMS GS (Ground Segment) consists of four GCs (Ground Centers); Satellite Control Center (SOC), National Meteorological Satellite Center (NMSC), Korea Ocean Satellite Center (KOSC), and Communication Test Earth Station (CTES) (KARI, 2006).

The SOC performs the primary satellite operation/monitoring and the secondary image data processing. The NMSC and KOSC have a role of the primary image data processing for MI (in NMSC) and GOCI (in KOSC), respectively, and The NMSC is also the secondary ground center for a satellite operation/monitoring. The CTES monitors RF (Radio Frequency) signals to check the status of Ka-Band communication system.

The SOC has two functions of the COMS GS; MI/GOCI Image data processing (as the backup center) and satellite operation/monitoring (as the primary center). One of SOC function is implemented in IDACS (Image Data Acquisition and Control System) for Image data processing by three subsystem; DATS (Data Acquisition and Transmission Subsystem), IMPS (IMage Pre-processing Subsystem), and LHGS (LRIT/HRIT Generation Subsystem) (Lim et al., 2011).

The other SOC function, satellite operation and monitoring, is implemented in SGCS (Satellite Ground Control System) by five subsystems; MPS (Mission Planning Subsystem), TTC (Telemetry, Tracking, and Command), ROS (Real-time Operations Subsystem), FDS (Flight Dynamics Subsystem), and CSS (COMS Simulator Subsystem) (Lee et al., 2006).

Fig. 12 shows the essential architecture of COMS ground segment with key composing subsystems and Table 3 describes functions of subsystem for COMS ground segment; DATS, IMPS, LHGS (IDACS) MPS, TTC, ROS, FDS, and CSS (SGCS).

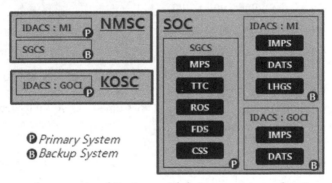

Fig. 12. COMS ground segment architecture with key composing subsystems

System	Sub-System	Functions
IDACS	DATS	Reception and error correction of CADU Processing and dissemination of LRIT/HRIT Control and monitoring of IDACS
	IMPS	CADU receiving and processing Radiometric correction (IRCM) Geometric calibration (INRSM) Payload status monitoring Interfaces among subsystems of the IDACS
	LHGS	LRIT/HRIT generation Compression and encryption for LRIT/HRIT generation

System	Sub-System	Functions
SGCS	MPS	Mission request gathering Mission scheduling Mission schedule reporting
	TTC	Telemetry reception Command transmission Tracking and ranging Control and monitoring
	ROS	Telemetry processing Telemetry analysis Command planning Telecommand processing
	FDS	Orbit Determination and prediction Station-keeping and re-location planning Satellite event prediction Satellite fuel accounting
	CSS	Satellite dynamic static simulation Command verification Anomaly simulation

Table 3. Functions of the COMS ground segment

3. COMS in-orbit performances

3.1 COMS AOCS performances and platform stability

The quality of images taken by on-board optical instruments is strongly dependent on the quality of the platform stabilisation. Three (3) strong requirements have been put on the COMS platform, all necessary to obtain the specified image quality.

- pointing accuracy (pitch and roll) : this specification is essential to a priori know where the instrument line of sight is aiming at. This is important for Ka band payload operations, for GOCI operation (due to further stitching of small images to construct the large imaging area) and for MI which can be commended to frequently review some local areas.
- pointing knowledge (pitch and roll) : the pointing knowledge is mainly driven by the INR in order to start the landmark matching processing with a sufficient accuracy.
- pointing stability (pitch and roll) : this specification is mainly driven by the GOCI instrument, requesting integration times as long as 8 seconds, with a jitter less than 10µrad.

The first point is fulfilled by the heritage bus (E3000 platform), but the two last points have necessitated the implementation of a high precision Fibre Optic Gyro (Astrium's FOG Astrix 120 HR), furthermore the third point has been flown down to micro-vibration dampers under wheels, various AOCS tuning (solar array natural mode damping, optimised wheel zero crossing management), optimized manoeuvres (reaction wheel off loading, EW and NS manoeuvres, etc.), and few operational constraints (stop solar array rotation during GOCI imaging period, etc.).

The resulting performances are typified as the pointing knowledge of better than 0.003°, the pointing accuracy of better than 0.05°, and the pointing stability of better than 7μrad/8s, all in roll and pitch. Fig. 13 shows the typical example of the performance on the platform stability.

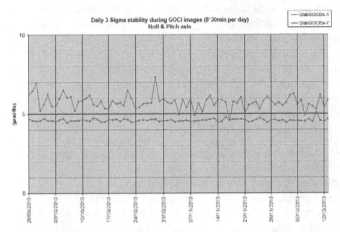

Fig. 13. COMS platform stability, as measured for a period of 3 months and computed on a 3-sigma basis

3.2 Radiometric performances of MI and GOCI

3.2.1 MI radiometric performances

3.2.1.1 MI in-orbit SNR

From the MI Visible dark noise analysis results, COMS MI in-orbit SNR at 5% albedo have been computed for both side 1 and side 2 of MI and the in-orbit SNR in both sides proved to be better than on ground measurement and significantly above the specification, SNR > 10 at 5% albedo. Table 4 shows both the on-ground and in-orbit SNR at 5% albedo for MI side 1.

MI Side 1	SNR 5%	
	On Ground	In Orbit
Detector 1	24	27.18
Detector 2	24	26.28
Detector 3	23	26.20
Detector 4	24	27.01
Detector 5	23	25.24
Detector 6	23	27.08
Detector 7	23	26.00
Detector 8	24	26.21

Table 4. MI On-Ground and In-Orbit SNR results, MI side 1

3.2.1.2 MI in-orbit radiometric calibration

COMS IOT (In-Orbit Test) MI calibration activities were divided into two main parts: MI visible channel and infrared channel calibrations. The visible channel calibration was conducted from July 11, 2010 after the COMS Launch (2010.6.26. 21:41 UTC). Calibration activity of the Infrared channels including the visible one was started from Aug 11, 2010 after the completion of the out-gassing (removal of remnant volatile contaminants by heating). The functional and performance tests were performed for both two functional sides (SIDE1: primary, SIDE2: secondary) plus two patch temperatures (patch Low and Mid) of the MI payload. In addition to the images of MI channels, albedo monitor and moon images were also acquired and analyzed. The final performance verification was checked officially at the phase 1 & phase 5 end meeting (Jan 26, 2011) after the intensive MI radiometric calibration processes conducted from July, 2010 to Jan, 2011. Summary of the verifications at the meeting is listed as follows.

1. Command and control tests for both sides (Side 1/Side2) were successful
2. Scan mechanism tests were successful
3. Image monitoring and acquisition tests were successful
4. The performance tests of MI visible channel were successful
5. The performance tests of MI infrared channels based on the payload real-time operational configuration modes were successful

3.2.1.2.1 MI visible channel calibration process

As shown in Fig. 14, the MI visible channel calibration process was simply verification of a linear visible calibration equation using the real data sets. After that, necessity of the normalization among eight detectors was checked. Albedo monitor data analysis and moon image processing were used for the detector's trend monitoring.

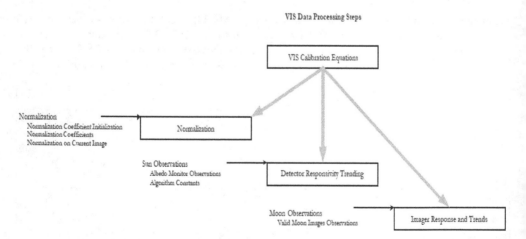

Fig. 14. MI visible channel radiometric calibration process flow chart

The pixel-to-pixel response non-uniformity (PRNU) were examined using the both space look and image data (Fig. 15). PRNU met the requirement specifications (denoted by red

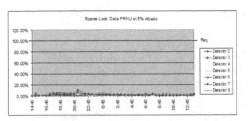

Fig. 15. PRNU Check (SIDE1): Space-Look data

lines). As a result, the normalization algorithm was not implemented on the visible channel calibration process.

3.2.1.2.2 MI infrared channel radiometric calibration

Different from the visible channel, the MI infrared channel calibration process has complex steps to get qualified data as shown in Fig. 16. First, coefficients of the basic (nominal) IR calibration equation were verified using the real data sets and then four major steps were taken: 1) Scan mirror emissivity compensation, 2) Midnight effect correction, 3) Slope averaging and 4) 1/f noise compensation.

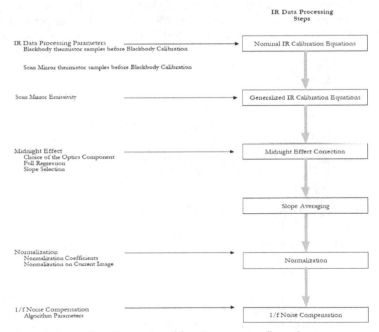

Fig. 16. MI infrared channel radiometric calibration process flow chart

1. Scan mirror emissivity correction

Based on the scan mirror emissivity (as a function of a scan angle), the effect of emitted radiances from the coating material on the scan mirror were compensated. The computed scan mirror emissivities according to different scan angles are shown in Fig. 17.

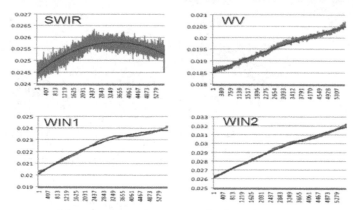

Fig. 17. Computation of the scan mirror emissivity for four different infrared chennels (1Dark Image, Side1, Patch Low, Det A, 2010.8.16).

2. Midnight effect compensation

Before and after four hours of near local midnight data were corrected using a mid night compensation algorithm (see Fig. 18). The estimated slope(open circles and squares) based on the regression between the black body slope and the selected optic temperature were used near midnight and the original slope values(thick lines) are used during the rest of time.

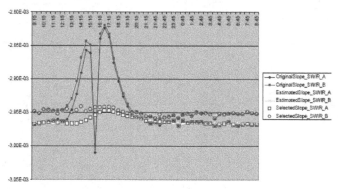

Fig. 18. IR Midnight Effect: the result of slope selection (SWIR; Side 1/Patch Low)

3. Slope averaging

Slope averaging is a smoothing process to remove the responsivity variation of the detectors due to the diurnal variation of background radiation inside the sensor. The reference slope value were compared to that of the previous day and the residual between two were filtered by the slope averaging.

4. 1/f noise compensation

The 1/f noise compensation, which is a filtering of random noise on the lower frequency components was also conducted. After the 1/f noise compensation, the stripping effects on the water vapor channel were greatly removed.

3.2.1.2.3 The result of the MI IOT radiometric calibration processes

The PRNU values from the radiometric indices computed from the real time MI data processing system of COMS (called IMPS) indicated that relative bias between detectors of infrared channels were minimal and thus the normalization process step on the infrared channels were skipped as same as the visible one. The complete COMS MI images resulted from the IOT (see Fig. 19) showed that the radiometric performance of the MI payload meets the all requirement specifications for the current operation configuration of MI (SIDE 1, Patch Low).

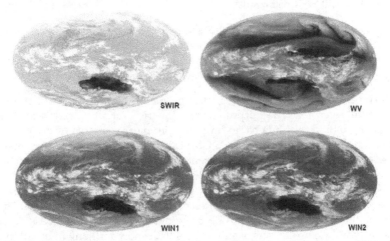

Fig. 19. Calibrated MI FD Level 1A images (before INR), (Side 1, Patch Low, 2010.12.23)

3.2.2 GOCI radiometric performances

3.2.2.1 GOCI in-orbit SNR

The GOCI was turned on for the first time in orbit on July 12, 2010 and captured it first image the day after. Both sides (primary and redundant) were successfully tested during about two weeks. After the successful functional tests such as the mechanism movement, detector temperature control, and imaging chain validity, the radiometric performance tests and radiometric calibration tests have been performed. The radiometric performance test is aimed to verify the validity of performance measured on ground. In-orbit offset and dark signal shows a quite good correlation with the ground measurements. Also the radiometric gain matrix, which has been measured in-orbit, is very similar to the ground gain. The SNR test results, which are provided in Table 5, show the performance exceeding the requirements in all 8 spectral bands by 25 to 40%. This is mainly due to the excellent quality of the CMOS matrix detector, and the design margin considered for worst case analysis.

3.2.2.2 GOCI in-orbit radiometric calibration

GOCI in-orbit radiometric calibration relies on a full pupil Sun Diffuser (SD), made of fused silica, known to be insensitive to radiations. The instrument is designed to allow a calibration every day. In practice, during IOT, two calibrations per week were performed. After IOT, the frequency of calibration was reduced to one per week. The

Band	Mean SNR	SNR specification at GOCI level
B1	1476	1077
B2	1496	1199
B3	1716	1316
B4	1722	1223
B5	1586	1192
B6	1513	1093
B7	1449	1107
B8	1390	1009

Table 5. GOCI In-Orbit SNR test result

potential aging of the SD is monitored by a second diffuser (Diffuser Aging Monitoring Device: DAMD) used less frequency than the SD, typically once per month since the end of the IOT. When not in used, both SD and DAMD are well protected by the shutter wheel cover to minimise their exposure to the space environment.

Through IOT period, about six months, the instrument calibration and the calibration stability were fully verified. The purpose of radiometric calibration test is to verify the in-orbit calibration method which is based on two point measurements (Kang & Coste, 2010). The in-orbit radiometric gain matrix of GOCI is calculated by using two sun images, which are obtained through the SD with two different integration times. The imaging time for the sun has been specified according to the desired solar incident angle over 25 degree to 35 degree. The actual solar incident angle of measured sun image is calculated by using the On-Board Time (OBT) which is included in the secondary header of the raw data. During IOT, sun imaging for eight spectral bands has been performed over two days based on one week period. For each calibration, six sets of sun images with short and long integration time have been obtained for each spectral band over about 10 minutes. Variation of gains calculated by 6 sets are very small (0.1 % to 0.3%) and are most probably due to processing noise (small errors in the ephemerides and in the calibration time) and also possibly to short term variations of the sun irradiance. Fig. 8 shows the gain evolution over eight months. For first three months, the gain shows a relatively rapid decrement. There is about 2% variation over eight months. Fig. 20 shows the aging factor of the SD over eight months. The trend provided in this Figure shows a sinusoidal variation over 8 months with about maximum

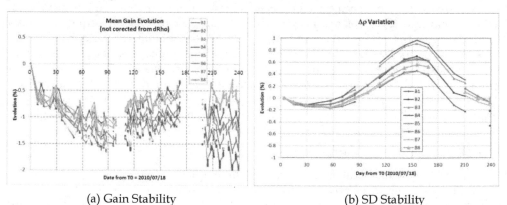

(a) Gain Stability   (b) SD Stability

Fig. 20. In-orbit radiometric stability over 8 months

1% amplitude. This is probably not the real variation of SD. In addition the longitudinal solar incident angle to the GOCI shows the similar variation over the year. The reason for this sinusoidal variation is now under examination. The variations lower than 1% over almost one year shows the SD stability. All the variations observed in orbit up to now are within 1 to 2% which is very low and very satisfactory. Some evolutions seem to be correlated with the longitudinal solar incident angle. This opens the way to further improvement of the calibration model if necessary.

The major performances (Modulation Transfer Function – MTF and Signal to Noise Ratio – SNR) are presented in this chapter, all other performances being well within the requirements.

One of the major advantages of ocean observation with the GOCI is that continuous monitoring is possible with images provided every hour, which maximizes chance of clear observation of the whole field even in cloudy season. No sun glint occurs thanks to the angular position of the field of view during daytime, while it discards many observations in low orbit.

3.3 Spatial and geometric performances of MI and GOCI

3.3.1 GSD and MTF

3.3.1.1 MI

During IOT, the MI Ground Sampling Distance (GSD) and the spatial performance (MTF) have been fully checked and verified. The GSD has been verified as follows. The landmark matching results by the INRSM were used and the angular steps in both E/W and N/S were measured by best fit between level 1A image coordinates and landmark GEOS positions. Those angular steps were used to determine a projection function for each image (or sub-image). Then, the specified GSD at Nadir was verified using the relevant projection function.

Table 6 shows the measured MI MTF results.

MTF	Side 1		Side 2		Required Minimum
	EW	NS	EW	NS	
0.25 Nyquist normalized wrt IFOV, i.e. 28µrad	0.93	0.98	0.92	0.98	0.87
0.50 Nyquist normalized wrt IFOV, i.e. 28µrad	0.83	0.92	0.83	0.92	0.68
0.75 Nyquist normalized wrt IFOV, i.e. 28µrad	0.68	0.74	0.69	0.72	0.49
1.00 Nyquist normalized wrt IFOV, i.e. 28µrad	0.52	0.45	0.54	0.47	0.20

Table 6. Measured In-Orbit MI MTF

3.3.1.2 GOCI

During IOT, the GOCI GSD has been verified by the same method as with MI, and the imaging coverage and the slot overlap have also been fully verified. The spatial performance (MTF) has also been checked. Before launch, the GOCI MTF performance was tested through ground test at the payload level. The in-orbit MTF test would allow the validation of MTF at system level including the satellite stability performance. But the measurement accuracy for in-orbit test is much worse than the ground test depending on the availability and the quality of the transition patterns between bright and dark in the image. The GOCI

MTF is calculated by using the image having a radiometric transition (such as a coast line) which is equivalent to Knife Edge Function (KEF) measurement. Table 7 shows the GOCI MTF test result. Significant margins are demonstrated with respect to specifications; similar margins are present in all spectral bands.

MTF @ Nyquist	Band 8			
	E	W	N	S
Sample #1	0.34	0.42	0.38	0.30
Sample #2	0.27	0.43	0.37	0.36
Sample #3	0.29	0.33	0.42	0.33
Sample #4	0.28	0.45	0.37	0.32
Sample #5	0.35	0.37	0.37	0.26
Sample #6	0.40	0.42	0.38	0.31
Sample #7	0.31	0.44		
Sample #8	0.43	0.34		
Sample #9	0.32	0.28		
Sample #10	0.32	0.36		
Mean Value	0.36		0.35	
Standard Error	0.06		0.04	
Specification	0.30		0.30	
(Mean - Spec.) / Spec.	19%		16%	

Table 7. Measured GOCI MTF in the band 8

3.3.2 INR performances

The INR IOT took a significant amount of time, as the final tuning requested. The first positive result obtained from the first images was the number of landmarks automatically extracted by the INR software. During the development, it had been demonstrated that a minimum of typically 100 landmarks were necessary, and sometimes more than 600 landmarks could be found on images.

The INR performance is evaluated on the basis on land marks residuals (statistical error after landmark best fit). In order to verify the validity of this approach, the coast line from the images is checked against an absolute coast line (based on GSHHS). The following figures in the Table 8 illustrate the typical performances of COMS INR as observed during the IOT.

	Navigation		Within Frame Reg.		Registration 15-min		Registration 90-min	
	EW	NS	EW	NS	EW	NS	EW	NS
Specification	56.0	56.0	42.0	42.0	28.0	28.0	42.0	42.0
Spec + Allocation VIS only	65.3	65.3	63.4	63.4	55.2	55.2	63.4	63.4
Jan 2011, VIS only	43.0	35.6	52.1	46.2	26.8	23.4	27.4	24.5
Sep 2010, VIS only	31.8	30.5	39.5	42.7	16.0	17.5	19.2	19.5
Spec + Allocation VIS & IR	87.5	87.5	103.9	103.9	99.1	99.1	103.9	103.9
Jan 2011, VIS & IR	46.3	43.7	58.8	55.8	39.8	37.7	33.4	33.2
Sep 2010, VIS & IR	40.5	40.6	50.6	54.1	23.9	23.8	27.1	27.4

	Navigation		Within Frame Reg.		Frame-to-Frame		Band-to-Band	
	EW	NS	EW	NS	EW	NS	EW	NS
Specification	28.0	28.0	28.0	28.0	28.0	28.0	7.0	7.0
Spec + Allocation VIS only	31.3	31.3	34.3	34.3	34.3	34.3	21.0	21.0
Jan 2011	17.5	15.5	22.9	21.2	7.7	7.3	<10.0	<9.6

Table 8. COMS MI and GOCI INR performances (units in μrad)

Worth noting is the fact that the COMS AOCS pointing performaces, as described in section 3.1, provide a significant contribution to the final INR performances. Also worth noting is the timeliness requirement put on the MI INR processing. As mentioned in section 2.2, the satellite serves as telecommunication relay to broadcast corrected data to end users in international formats called HRIT and LRIT. Both formats suppose to rectify the data both radiometrically and geometrically. An allowance of 15 minutes is given to perform the ground processing before uploading again the data to the satellite. After few inevitable tunings, the whole process is now performed in typically 12 minutes. For illustration purpose, two examples of shoreline matching are presented for MI vis channel and for one GOCI spectral band in the Fig. 21 and Fig. 22.

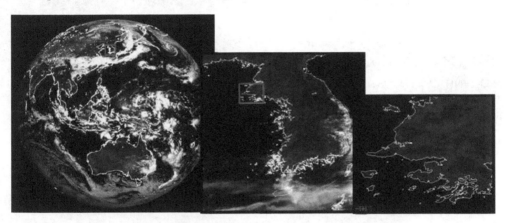

Fig. 21. MI shoreline matching (FD, VIS)

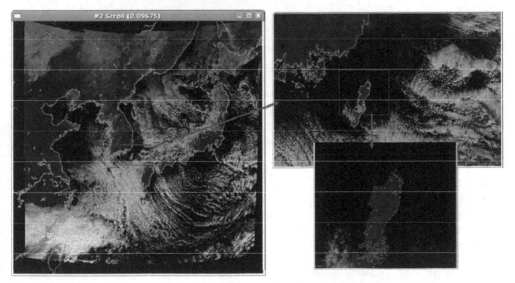

Fig. 22. GOCI shoreline matching. The reference shoreline is superimposed to the geometrically rectified GOCI image. Matching is better than 2 pixels over the whole area.

Further analysis and monitoring on INR performances have been performed since the start of normal operation of COMS for the service to the end users, and Fig. 23 and Fig. 24 illustrate some of these typical COMS INR performances.

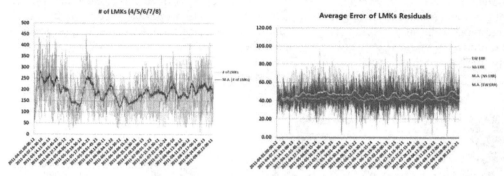

Mode: ENH, Channel: VIS and IR, and negative correlation between the number of LMKs and the average of Residuals (courtesy of KMA)

Fig. 23. MI INR performance during 1st April ~ 31th August.

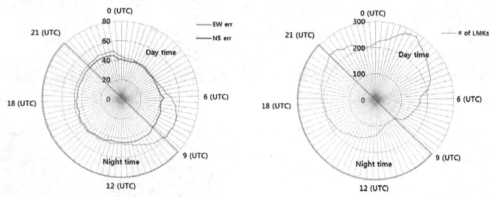

Mode: ENH, Channel: VIS and IR, and negative correlation between the number of LMKs and the average of Residuals: Twilight effects (courtesy of KMA)

Fig. 24. MI INR performance during 1st April ~ 31th August.

4. Application and suggestion

It has been merely 8 months since the outset of the normal operation of COMS for the distribution and service of the images and image products to the end users and scientific communities. The activities in this period in terms of the data processing, calibration and the end product generation and the related studies and researches have been exceedingly interesting, proactive and imaginative, to say the least, and in a word 'dynamic' in a very positive and rewarding sense. This section describes the application aspect of the COMS image data from MI and GOCI, addresses some posing technical challenges at the present time on this course of data application, summarizes some of the representative end products both from MI and GOCI and discusses the way forwards with some suggestions.

4.1 MI

4.1.1 Generation of MI end products

As mentioned in the previous sections, COMS MI Level 1B data are generated through radiometric and geometric calibrations and then sixteen meteorological products(level 2) are produced by CMDPS (COMS Meteorological Data Processing System) as shown Fig. 25.

Fig. 25. COMS Meteorological Products

Parts of meteorological products from COMS MI have been generated operationally since April 1, 2011 together with COMS operation. Those products are cloud analysis (type, phase and amount), cloud top temperature/pressure, atmospheric motion vector, cloud detection, fog, and aerosol index. And then, four products, which are sea surface temperature, rain

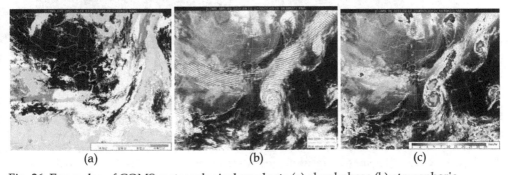

Fig. 26. Examples of COMS meteorological products (a) cloud phase (b) atmospheric meteorological vector and (c) rain intensity.

intensity, outgoing longwave radiation, and upper tropospheric humidity, were generated additionally from 10 August 2011. These products are currently being validated through comparison between satellite-derived products and ground in-situ data. For example, detection area of Asian dust (aerosol index) occurred in 2011 April and May was compared with COMS GOCI and MODIS (Moderate Resolution Imaging Spectroradiometer) true color images or OMI (Ozone Monitoring Instrument) AOD (Aerosol Optical Depth). The other six products which are land surface temperature, sea ice/snow cover, total precipitable water, insolation, clear sky radiance, and aerosol optical depth will be operationally produced soon.

4.1.2 Application to weather forecasting and analysis

In Korean peninsula, annual losses and damages in human and material are enormous due to the convective cloud accompanying summer heavy rainfall, which is either flown from the West Sea or originated locally. COMS can monitor and watch the origination and development of this convective cloud since it can observe Korean peninsula with MI in a concentrative way eight times an hour. NMSC is supporting the weather forecasting with the developed technique for Very Short Range Forecasting utilizing COMS MI meteorological data, which was introduced and derived from the technique of convective cloud rainfall intensity calculation and monitoring by the SAFNWC (Satellite Application Facilities Nowcasting) of EUMETSAT (European Organization for the Exploitation of Meteorological Satellites).

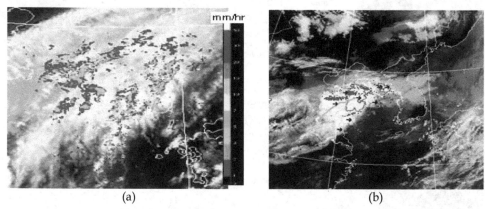

(a)   (b)

Fig. 27. Examples of COMS MI data applications (a) Convective rain intensity image combined with radar rain map (b) Predicted location of convective cloud and lightening image.

To analyze the Typhoon, which passes through Korean peninsula two to three times a year, typically around July to September time, such elements as the Typhoon intensity, radius of strong winds, the maximum wind speed, low pressure, are needed. In this analysis, NMSC is utilizing the Advanced Dvorak Technique (ADT) in the site operations, which was developed by the Cooperative Institute for Meteorological Satellite Studies (CIMSS) of University of Wisconsin (UW). The algorithm in this technique classifies the evolution phase of the tropical cyclone according to its intensity, as the formation phase, the development phase and the disappearance phase, based on the MI infrared (IR) images, and automatically analyses the Typhoon intensity through the experience in pattern recognition by applying the Fast Fourier Transform (FFT) on the resulting patterns from the different phase of the cyclone.

COMS MI data are also to be used in the generation of aeronautical meteorological products, as shown in the Fig. 28. These products may have a relatively low accuracy but have the advantage of observing the broader area every hour. They are providing the level 2 information; such as the cloud phase, cloud height and the cloud top temperature in the air route and also the information from the convective cloud monitoring, and the other technique is under development for the information generation on the elements that can cause aircraft accidents, such as the icing and the turbulence.

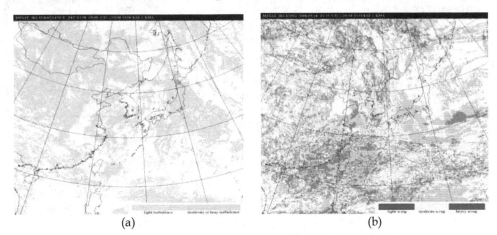

(a)   (b)

Fig. 28. Examples of COMS MI aeronautical meteorological products (under developments) (a) turbulence distribution (b) icing on airplane area .

4.2 GOCI

The application of GOCI data is focused on the monitoring of long-term/short-term ocean change phenomena around Korean peninsula and north-eastern Asian seas. In daytime, the hourly-produced GOCI data will be used for the ocean/coast environmental monitoring and for the observation of ocean dynamics features and the management of ocean territory. Also, these GOCI data, when used in conjunction with ocean numeric models, would bring forth the increase of accuracy in ocean forecasting,.

GOCI level 2 data products can be generated from GOCI level 1B with GDPS (GOCI Data Processing System) which is the data processing and analysis software developed by KORDI.

This GDPS system derives the pure ocean signal (water leaving radiance) by atmospheric correction using aero-optics model and oceano-optics model developed and modified by KORDI. It can extract pure water signal as the normalized water leaving radiance which is corrected water leaving radiance by considering the satellite - sun relative geometry. For geostationary satellite, this relative position of the sun and the satellite changes all the time and then the ocean signal is distorted. To resolve this issue of signal distortion, some research was performed. The system can generate the marine environment analysis data using specific algorithms for target region. The data processing algorithms applied to the existing ocean satellite optical sensor and new algorithms to the GOCI would produce the latest marine environmental analysis results.

Table 9 shows the list of GOCI level 2 data products which are currently being generated and used for each application purpose, and Table 10 signifies the list of GOCI level 3 data products which can also be generated by GDPS. The algorithm to generate GOCI level 3 data is under the final validation process. Fig. 29 shows some typical examples of these end products, in the case of TSS and CDOM.

PRODUCTS	DESCRIPTION	APPLICATION
Water-leaving Radiance (Lw)	The radiance assumed to be measured at the very surface of the water under the atmosphere	Indispensible for water color analysis algorithms
Normalized water leaving radiance (nLw)	The water leaving radiance assumed to be measured at nadir, as if there was no atmosphere with the Sun at zenith	Input data for the water analysis algorithm
Chlorophyll (CHL)	Concentration of phytoplankton chlorophyll in ocean water	Ocean primary production estimation, dumping site monitoring, climate change monitoring
TSS	Total suspended sediment concentration in ocean water	Coastal ocean environmental analysis and monitoring TSS movement and transfer monitoring
CDOM	Colored dissolved organic matter concentration in ocean water	Indicator of ocean pollution Ocean salinity estimation
Optical properties of water	K-coefficient Absorption coefficient(a) Backscattering coefficient(bb)	Ocean optical properties analysis
Red tide (RI)	Red tide index information	Ocean pollution and ecological monitoring Movement and transfer monitoring of red tide
Underwater Visibility (VIS)	Degree of clarity of the ocean observed by the naked eye	Navy tactics, ocean pollution map, sea rescue work
Atm. & earth environment	Yellow dust, Vegetation Index	Atmospheric environment and land application

Table 9. GOCI level 2 data products

PRODUCTS	DESCRIPTION	APPLICATION
Daily composite of CHL, SS, CDOM	Daily 8 images composite for cloud free mosaic image	Climate change trend analysis
Fishing ground Information (FGI)	Fishing ground probability index, fishing ground prediction	Fishing ground detection Fishing ground environmental information

PRODUCTS	DESCRIPTION	APPLICATION
Sea surface current vector (WCV)	Sea surface current direction/speed	Understanding of sea surface currents and estimation of pollutant movements
Water quality Level (WQL)	Coastal water quality level estimation	Coastal ocean eutrophication Coastal water quality control/monitoring
Primary Productivity (PP)	The production of Organic compounds from carbon dioxide, principally through the process of photosynthesis	Carbon cycle Long-term climate change monitoring

Table 10. GOCI level 3 data products

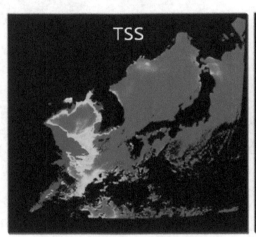

Fig. 29. Examples of GOCI level 2 end products, TSS (Total Suspended Sediment) and CDOM (Colored Dissolved Organic Matter)

GOCI products such as ocean current vector and ocean color properties would be provided to the fishery and the related organization for the increase of the haul, the effective management of fish, and finally the increase of fisheries income. The GOCI data could also be useful for monitoring suspended sediment movement, pollution particles movement, ocean current circulation and ocean ecosystem. Also, it will contribute to the international cooperation system, such as GEOSS (Global Earth Observation System of Systems), for the long-term ocean climate change related research and application by the data exchange and co-research among related countries.

The Korea Ocean Satellite Center (KOSC) in KORDI as the official GOCI operation agency, receives the GOCI data from the satellite directly, generates, stores, manages and distributes the processed standard products. And KOSC will continuously develop new ocean environmental analysis algorithms to apply to the imagery data of GOCI and the GOCI-II which is next generation of GOCI.

Through the normal operation of GOCI, KOSC can provide the new, high-grade ocean environmental information in near-real time. It can be applied to the detection of freak phenomena of ocean nature such as the red-tide and the green-tide. The primary productivity derived from the GOCI chlorophyll and other products is the key research information about ocean carbon circulation. The color RGB images and analysis images of GOCI products with high spatial resolution are clearer and more recognizable than the monochrome images from other existing geostationary earth monitoring satellite which has only 1 visible band. These images can be useful to land application and atmospheric remote sensing application like monitoring of typhoon, sea ice, forest fire, yellow dust, etc.

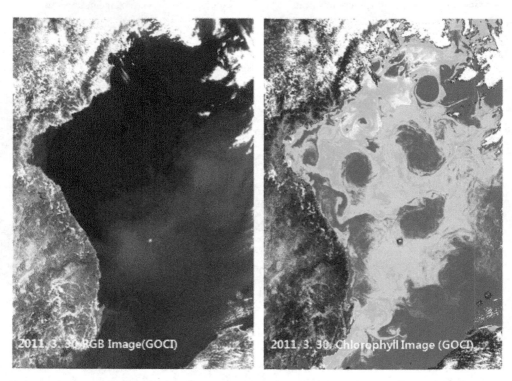

Fig. 30. Standard RGB image of GOCI (left) and the analysis result of seawater chlorophyll density in the East Sea (right)

Table 11 shows the overall scheme of GOCI data application, and Fig. 31 and Fig. 32 exemplify some of the typical applications. In Fig. 32, several Images of different dates were mosaiced to realize this cloud-free picture and the numerical signals of the Yellow Sea (East China Sea), the East Sea (Japan Sea) and Northwestern Pacific were differently processed to maintain a balanced tone throughout the whole coverage area of the GOCI.

Operation	Application Items
Carbon Circulation Monitoring	- Analysis of marine primary productivity - Long-term climate change research in the ocean - Long-term analysis of climate change through research studies and utilized to secure carbon credits

Operation	Application Items
Red Tide Monitoring	- Amounts of red tide, forecast to move through the path and spread of red tide-related damage contribute to the reduction
Green Algae Monitoring	- Amounts of green algae, forecast to move through the path and spread of green algae-related damage contribute to the reduction
Oil spill monitoring	- Oil spill and monitoring of movement and distribution of pollution
Speculative waters, environmental monitoring	- Ocean dumping in the waters of chlorophyll contained in phytoplankton concentration and monitoring the amount of organic matter dissolved in seawater
Turbidity Monitoring	- Indicators of marine pollution - The total suspended inorganic material contained in seawater through the coastal marine environment observation analysis and monitoring
Low-salinity water monitoring	- Utilization of seawater salinity estimates (Low-salted water monitoring) - Which flows from China to determine the utilization of contaminant migration path
Fishery Information	- Fish and Fishery distribution - Fisheries and environmental monitoring and fisheries contribute to productivity improvement
Fisheries and fish-farm management	- Long-term monitoring of marine ecosystems through the efficient management of fisheries resources
Ecological monitoring tidal	- Marine Biology / Ecological Survey, appeal / river forecasts, and productivity of aquatic organisms in the environment - Coastal fisheries resources management
Hurricane watch	- A hurricane tracking and navigation path - The impact of Hurricane directed by the ocean, producing flow information
Sea-ice monitoring	- Development of the area of sea ice observations and monitoring - Support fishing operations
Forest fire monitoring	- land management and forest fire monitoring, forest resources utilization
Dust monitoring	- Dust, vegetation, and the atmosphere and global environmental monitoring information - Dust weather analysis and forecasting, and utilized in the atmospheric environment
Current surveillance	- Balm of seawater, and the flow rate information production - Utilize coastal water quality management
El Niño, La Niña monitoring	- Estimated using ocean temperature and productivity monitoring long-term climate change

Table 11. Application subjects of GOCI data

Fig. 31. Land and Sea Features expressed by natural color on the full scene of the GOCI.

Fig. 32. Structure of Chlorophyll Distribution in the North-East Asian Seas.

5. Conclusion

COMS is a unique bird in many ways, partially in that it is such a complex satellite accommodating three different payloads with rather conflicting missions into a single spacecraft bus, partially in that it employs a unique and novel INR system, and also partially because it has the GOCI on board, the world's 1st geostationary imager for the ocean colour. By the joint effort of EADS Astrium and KARI, it was masterfully designed, developed, tested, and launched, and is now behaving beautifully in orbit, exhibiting quite impressive and fruitful performances along with the very useful and interesting image data and processed end products.

It is especially interesting to note that with the co-existence of both MI and GOCI on board, the comparison and combination of data taken by these two sensors from the same geostationary location, could open some new windows for further interesting research and development. In the case of GOCI, the benefits of Geo observation compared to its LEO (Low Earth Orbit) counterparts has been notably demonstrated and largely appreciated by the end users so far, even with the relatively short accumulated time of normal service, and as the further activities on post processing and related studies will get refined and matured, it is expected that this trend will become even more prominent.

With these observations, findings and expectations at hand, it could be cautiously said that COMS image data and the processed end products will bring some added dimension to the world remote sensing community and the related field of science and technology. For this end, it will be made sure that the application of MI and GOCI data during the mission life of COMS is to be fully exploited and maximized. It is hoped and believed that all the aspects of the COMS development and operation; from the design, implementation, test and validation, launch and IOT, to the data processing, end product generation, data utilization and end user services will continue to grow and be improved and expanded in its relevant realm into the next generation of geostationary remote sensing satellites.

6. Acknowledgment

COMS program has involved so many different organizations, agencies, government bureaus and companies and wide spectrum of participating personnel with different cultures, characters and backgrounds. It is with such a great emotion and gratitude, along with the highly rewarding feeling and sense of proud accomplishment, that we can now say that we regard all the participating members in this program as one big 'COMS family' and to some, very close life-long friends, indeed. There were certainly some bumpy roads and rocky times in the course, but through them all we became real friends and it is grateful that we can now look back upon those days with sense of mutual respect and appreciation.

We feel deeply grateful and obliged to send our appreciation to our Korean government bureaus first and foremost; MEST (Ministry of Education, Science and Technology), KMA (Korea Meteorological Administration), KORDI (Korea Ocean Research & Development Institute), MLTM (Ministry of Land, Transport and Maritime Affairs) and MIC (Ministry of Information and Communication), among others, without whose support and dedication this grand program would not have been possible. We feel especially thankful to MOSF (Ministry Of Strategy and Finance) for providing us the actual revenue sources continually throughout the entire course of this challenging program.

The author and co-authors of this chapter only represent a very small portion of all COMS family members, and we believe that the authors of this chapter, in fact, ought to be all COMS family members and thus we feel deeply indebted to them. Our special thanks go to; Mr. Seong-rae Jung and Ms. Jin Woo of KMA, and Mr. Hee-jeong Han and Mr. Seong-ik Cho of KORDI, for the charts in the section 3.3.2 and for their great help and support in preparing and finalizing sections 4.1 and 4.2.

Last but clearly not the least, we remember our missing COMS family members, Mr. Daniel Buvat of EADS Astrium and Mr. Young-joon Chang and Mr. Sang-mu Moon of KARI, who abruptly departed from this life on earth in the course of COMS development and operation, leaving the rest of us in deep grief and helpless devastation. Along the lines of COMS history, with the trace of their sincere commitment and contribution to the success of COMS, they will always be remembered in our hearts. We dedicate this small chapter to them.

7. References

Cros, G.; Loubières, P.; Lainé, I.; Ferrand, S.; Buret, T.; Guay, P. (June 2011) *European ASTRIX FOGS In-Orbit Heritage*, 8th International ESA Conference on Guidance, Navigation & Control Systems

Kang, G.; Coste, P. (2010) *An In-orbit Radiometric Calibration Method of the Geostationary Ocean Color Imager (GOCI)*, IEEE Transactions on Geoscience and Remote Sensing, Digital Object Identifier 10.1109 / TGRS. 2010. 2050329

Kim, H.; Kang, G.; Ellis, B.; Nam, M.; Youn, H.; Faure, F.; Coste, P.; Servin, P. (2009) *Geostationary Ocean Color Imager (GOCI), Overview and Prospect*, 60th International Astronautical Congress (IAC 2009)

Kim, H.; Meyer, P.; Crombez, V.; Harris, J. (2010) *COMS INR: Prospect and Retrospect*, 61st International Astronautical Congress (IAC 2010)

Lambert, H.; Koeck, C.; Kim, H.; Degremont, J.; Laine, I. (2011) *One Year into the Success of the COMS Mission*, 62nd International Astronautical Congress (IAC 2011)

KARI (Korea Aerospace Research Institute) (January 2006). *COMS Ground Segment Specification, Ref C1-SP-800-001-Rev.C*, Deajeon, KOREA

Lee, B.; Jeong, W.; Lee, S.; et al. (April 2006). *Funtional Design of COMS Satellite Ground Control System*, Conference of the Korean soceiety for aeronautical and space sceience, pp. 1000-1005, KSAS06-1850.

Lim, H.; Ahn, S.; Seo, S.; Park, D. (December 2011). *In-Orbit Test Operational Validation of the COMS Image Data Acquisition and Control System*, Journal of the Korean soceiety of Space Tehnology, Vol.6 No.2, pp. 1- 9.

2

Active Remote Sensing: Lidar SNR Improvements

Yasser Hassebo
LaGuardia Community College of the City University of New York
USA

1. Introduction

RAdio Detection **A**nd Ranging (RADAR), **SO**und **NA**vigation and Ranging (SONAR), and **LI**ght Detection **A**nd Ranging (LIDAR) are active remote sensing systems used for earth observations (Planes and ships' locations and velocity information, air traffic control, oceanographic and land info,), bathymetric mapping (e.g., hypsometry, Ocean depth (echo-sounding), SHOALS, and seafloor), and topographic mapping. Integrating laser with RADAR techniques – laser RADAR or LIDAR - after World War II introduces scientists to a new era of Remote Sensing technologies. LIDAR is one of the most widely used active remote sensing systems to attain elevation information which an essential component to obtain geographical data. While RADAR is transmitting a long-wavelength signal (i.e., radio or microwave: cm scale) to the atmosphere and then collecting the backscattering energy signal, LIDAR transmission is a short-wavelength laser beam(s) (i.e., nm scale) to the atmosphere and then detecting the backscattering light signal(s). More lidar principles and comparison between active remote sensing techniques are introduced in section 1.1 of this chapter.

2. Lidar background

2.1 Lidar historical background

After World War II the first **LI**ght Detection **A**nd **R**anging (lidar) system was invented (Jones 1949). The light source was a flash light between aluminum electrodes with high voltage amplitude transmitter, and the receiver optics were two mirrors. Afterward a photoelectrical cell was used as a detector. During daylight, this system had been used to measure the height of cloud ceiling up to 5.5 km. At that time the acronym *lidar* didn't exit (Middleton 1953). The real revolution of lidar began with the invention of the laser (light amplification by stimulated emission of radiation) in 1960. Using laser as a source of light in a lidar system is referred to as *"Laser Radar, or Ladar, or Lidar"*. Lidar operates in wide band region of the electromagnetic spectrum; ultraviolet (225 nm- 400 nm), visible (400 nm-700 nm), and infrared radiation (700 nm- 1200 nm). Lidar systems are used as ground based stations (stationary or mobile), or can be carried on platforms such as airplanes or balloons (in-situ operations), or on satellites. National Oceanic and Atmospheric Agency (NOAA) and National Aeronautics and Space Administration (NASA) aircraft and satellites are the most famous lidar platforms in the United States of America. Some other platforms are employed around the world by groups such as the European Space Agency (ESA), the Japanese National Institute for Environmental Studies (NIES), and the National Space Development Agency of Japan (NASDA).

What is a lidar

LIght Detection And Ranging (lidar) is an optical remote sensing system for probing the earth's atmosphere through Laser transmitter using elastic and/or inelastic scattering techniques. Most of the remote sensing lidar systems consist of three functional subsystems, as shown in Figure 1, which vary in the details based on the particular applications. These subsystems are: (1) Transmission subsystem, (2) Receiver subsystem, (3) Electronics subsystem.

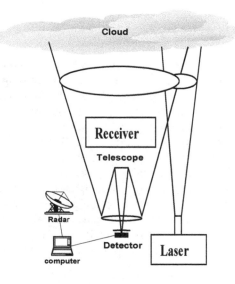

Fig. 1. Essential elements of a lidar system

In the **transmission** subsystem, a laser (pulsed or continuous wave (CW)) is used as a light source. More than one laser can be used according to lidar type and objective of the measurements. Laser pulses, in the ideal case, are very short pulses with narrow bandwidth, high repetition rate, very high peak power, and are propagated with a small degree of divergence. The laser pulse is transmitted through transmission optics to the atmospheric object of interest. The essential function of the output optics is to improve the output laser beam properties and/or control the outgoing beam polarization. Elements such as lenses and mirrors are used to improve the beam collimation. Beam expansion is used to reduce the beam divergence and the area density of the laser pulse. Fiber optic cable, filters, and cover shields or housings serve the dual purpose of preventing the receiver detectors from saturation due to any unwanted transmitted radiations and of protecting the user's eyes against any injury. Wave length selective devices are also used, such as harmonic generator, to create the second, the third and the fourth harmonic. Polarizer can be used to control the transmitted beam polarization. Polarization measurement equipments are used as well. The experimental results, in this chapter, had been produced using two types of pulsed laser, Q-

Switched (an optical on-off switch) Nd: YAG (Continuum Infinity 40-100) and Q-Switched Nd: YAG (Surelite) at CCNY.

A Receiver subsystem consists of an optical telescope to gather and focus the backscattering radiations, and receiver optics to provide the detector (PhotoMultiplier Tube (PMT) or Avalanche Photo Diode (APD)) with desirable collimated or/and focused strong polarized signal. Components such as mirrors, collimated lenses, aperture (field stop), ND (Neutral Density) filters, and Interference Filters (IF) are to provide special filtering against sky background radiations, analyzers (polarization selection components) are needed to select the necessary polarizations based on the applications and/or to discriminate against the unwanted background noise (as shown in chapter 7), and electro-optical elements that convert light energy to electrical energy (detectors). There are two basic types of detectors for lidar systems; the photomultiplier tube (PMT), and the avalanche photo diode (APD). In addition to the optics mounts and the manually operation aids, automated alignment capabilities for lidar long-term unattended operations are needed.

Electronic subsystem consists of data acquisition (mostly, multiple channels), displaying unites, Analog to Digital (A/D) signals conversions, radar and radar circuit, control system especially for our polarization discrimination technique which I presented in this dissertation (Chapter 7) to track the azimuth angles to improve the SNR. In addition, software (Labview and Matlab) is needed for signal processing purposes, as well some hardware such as platforms (van for a ground based mobile lidar, airplane or balloon for in-situ airborne lidar and satellite for higher altitude space-based scanning lidar), a temperature control unit, orientation stability elements, storage units and some additional equipment depending on lidar's type and measurement objective.

2.2 How does lidar work?

Using the well known fact that the laser energy of optical frequencies is highly monochromatic and coherent, and the revolution of developing the Q-Switching by McClung and Hellwarth on 1962, (McClung 1962), laser has the capability of producing pulses with very short duration, narrow bandwidth, very high peak energy, propagating into the atmosphere with small divergence degree. This prompted the development of backscattering techniques for environment and/or atmosphere compositions and structure, (aerosol, ozone, cloud plumes, smoke plumes, dust, water vapor and greenhouse gases (e.g. carbon dioxide), temperature profile, wind speed, gravity waves, etc.), distributions, concentrations and measurements. These measurement techniques are to some extent analogous to radar, except using light waves as an alternative to radio waves. Consequently, scientists denote lidar as laser radar. The essential idea of lidar operations and measurements is based on the shape of the detected backscattering lidar signals with wavelength of (λ) if the transmitted laser beam of wavelength (λ_L) is scattering back from distance R. This backscattering shape depends on the properties of the lidar characteristics and the atmosphere specifications. The transmitted lidar signal can be absorbed, scattered, and shifted, or its polarization can be changed by the atmosphere compositions and scattered in all the directions with some signals scattered back to the lidar receiver. Two parameters, in the lidar return equation, relate the lidar detected signal power and the atmospheric specifications. These parameters are the extinction and scattering coefficients, a (λ, R), β (λ, R) respectively. By solving the lidar equation for those coefficients one can determine various atmospheric properties. An example of these determination processes, which based on the lidar type and the physical process used in the measurements, have been introduced in this chapter.

3. Lidar classifications

Ways to classify lidar systems are: (1) the kind of physical processes (Rayleigh, Mie, elastic and inelastic backscattering, absorption, florescence , etc.), (2) the types of the laser employed (Die, and ND:YAG), (3) the objective of the lidar measurements (aerosols and cloud properties, temperature, ozone, humidity and water vapor, wind and turbulence, etc.), (4) the atmospheric parameters that can lidar measure (atmospheric density, gaseous pollutants, atmospheric temperature profiles), (5) the wavelength that been used in the measurements (ultraviolet (UV), infrared (IR), and visible), (6) the lidar configurations (monostatic, biaxial, coaxial, vertically pointed and scanning lidars and bi-static), (7) the measurement mode (analogue, digital), (8) the platform type (stationary in laboratories, mobiles in vehicles, in situ (balloon and aircraft), and satellite), and (9) number of wavelength (single, and multiple wavelengths). In the following section anticipate brief descriptions of various types of lidar, focusing mainly on those types of our research interest.

4. Types of lidar returns

If light is directed towards other directions because of interaction with matter without loss of energy (but losing intensity) the fundamental physical process is called *scattering* of light. The light scattering occurs at all wavelengths in the electromagnetic spectrum and in all directions. If lidars sense only the scattering radiations in the backward direction (scattering angle $\theta_s = 180°$ for monostatic vertically pointed lidar), we call them lidar *backscattering* radiations or signals. In terms of lidar return signals, lidar has been classified into the following types: Rayleigh, Mie, Raman, DIAL, Doppler, and florescence lidars.

4.1 Rayleigh scattering lidar

In 1871, Lord Rayleigh discovered a significant physical law of light scattering with a variety of applications. The most famous applications of this discovery are the blue sky and the sky light partial polarization explanations. Rayleigh scattering is elastic (no wavelength shift) scattering from atmospheric molecules (particle radius is much smaller compared with the incident radiation wavelength i.e. $r_p \ll \lambda$): sum of Cabannes (sum of coherent, isotropic, polarized scattering, which approximately 96% of the scattering) and rotational- Raman S and S′ branch scattering which only 4% of the scattering proceedings. Based on the Rayleigh-Jeans law, [the Planck radiance is linearly proportional to the temperature, $B(T) \approx (2\kappa_B v^2 / c^2)(T)$, where $B(T)$ is the Planck function, κ_B is the Boltzmann constant $\kappa_B = 103806 \times 10^{-16} erg.deg^{-1}$, v is the oscillator frequency, c the speed of light, and T the absolute temperature] (Liou 2002), Rayleigh lidar technique can be used to derive the atmospheric temperature profile above the aerosol free region ($R > 30\ km$). Since molecular scattering (Rayleigh scattering or aerosol-free scattering) is proportional to the atmospheric density, the atmospheric temperature profile can be simply derived from the atmospheric density in the range above the aerosol layers (above 30 km to below 80 km). Unfortunately, above 80 km temperature measurements require a powerful transmitter laser (up to 20 W) and receiver telescope (up to 4 m aperture) which are difficult for mobile or airborne platforms (Fhjii and Fukuchi 2005). Finally, assuming the atmosphere consists of molecules only and outside the gaseous absorption bands of the atmosphere, the atmosphere optical thickness can be approximated by

$$\tau_m = 0.008569\lambda^{-4}(1 + 0.0113\lambda^{-2} + 0.00013\lambda^{-4})$$, where λ is measured in micrometers.

Rayleigh scattering strongly depends on the wavelength of the transmitted light (λ^{-4}) which explains the blue color of the sky, where the scattering efficiency is proportional to λ^{-4}, i.e. rapid increase in the scattering efficiency with decreasing λ. This behavior leads to more scatter in blue than red light of the air molecules.

4.2 Mie backscatter lidars

For particle radius (r_p) larger than $\lambda / 2\pi$, (i.e., $r_p > \lambda / 2\pi$, where λ is wavelength of radiation), Rayleigh scattering is not applicable but Mie scattering applies (Mie 1908; Measures 1984; Liou 2002). Mie scattering is elastic scattering which is suitable for detection of large spherical and non-spherical aerosol and cloud particles mainly in the troposphere (Barber 1975), (Wiscombe 1980). The backscattering signals from aerosol or molecules and the absorption from molecules are very strong in the lower part of the atmosphere (below 30 km), which is enough to determine various properties about the atmosphere. Micrometer-sized aerosol and clouds are great indicators of atmosphere boundary phenomena where they show strong backscattering interaction. By Mie scattering theory, the optical properties of water droplets can be evaluated for any wavelength in the electromagnetic spectrum (from solar to microwave) (Deirmendjian 1969). Clouds covered about 50% of the earth (Liou 2002). Clouds also have an important impact on the global warming disaster when clouds trap the outgoing terrestrial radiation and produce a greenhouse gaseous effect. Mie backscattering lidar measures backscattered radiation from aerosol and cloud particles and their polarization as well (Mie 1908; Liou 2002). Its performance is similar to radar manner. A laser pulse of energy is transmitted, interacted with different objects and then backscattered (scattering angle =180°) to the receiver detector. The detected backscattering signals are interrelated with some properties of that object (even with low concentrations or small change in concentrations of dust or aerosol objects). Mie scattering follows $(\lambda^{-0}$ to $\lambda^{-2})$, i.e., it is not significantly dependent on the wavelength.

4.3 Raman (inelastic backscattering) lidars

Raman scattering is inelastic scattering with cross section up to three times smaller than the Rayleigh cross section in magnitude. A Raman scattered signal is shifted in frequency from the incident light (Raman-shifted frequency). The Raman scattering coefficient is proportional to the atmospheric density when the air molecule (nitrogen or oxygen) is used as Raman materials (Fhjii and Fukuchi 2005). Generally speaking, Raman lidar measure intensity at shifted wavelength (Stephens 1994) and it detects selected species by monitoring the wavelength-shifted molecular return produced by vibration Raman scattering from the chosen molecules. Raman lidar, originally, was developed for NASA Tropical Ozone Transport Experiment/ Vortex Ozone Transport Experiment (TOTE/VOTE) for methane (CH_4) and Ozone measurements (Heaps 1996). Also it has been used to correct the microwave temperature profile in the stratosphere (Heaps 1997).Typically; inelastic scattering (such as Raman) is very weak; therefore the daytime measurement is difficult due to the strong background solar radiation. This restricts Raman lidar measurements to nighttime use where background solar radiation is absent. On the other hand, Raman lidar is a powerful remote sensing tool used to measure and trace constituents where elastic lidar can not identify the gas species (Fhjii and Fukuchi 2005). Raman-Mie Lidar technique is also used to determine the extinction and the backscattering coefficients assuming the knowledge of air pressure (Ansmann 1992). In this chapter I introduced, a polarization technique to improve lidar Signal-to-Noise Ratio (SNR) by reducing the background noise during the daytime measurements. This will help for successful diurnal operation of Raman lidar.

4.4 DIfferential absorption lidar (DIAL)

Differential Absorption and Scattering (DAS) is a good combination for detecting a good resolution of water vapor in the atmosphere using the H_2O absorption line at 690 nm (Schotland 1966; Measures 1984). DAS technique is one of the best methods for detecting constituents for long-range monitoring based on a comparison between the atmospheric backscattering signals from two adjacent wavelengths that are absorbed differently by the gas of interest (Measures. R. M. 1972). The closest wavelength, of the two adjacent wavelengths, to the absorption line of the molecule of interest (i.e., strongly absorbing spectral location due to the presence of an absorbing gas) is usually called on-line and denoted as (λ_{ON}) and the other laser wavelength is called off-line and denoted as (λ_{OFF}). DIfferential Absorption Lidar (DIAL) technique is a unique method to measure and trace gaseous concentrations in the Planetry Boundary Layer (Welton, Campble et al.) (Welton, Campble et al.) in three dimensional mode (3D) using of the DAS principal. The gas number density $N_x(R)$ can be derived from the differential absorption cross section of the molecular species of interest ($\Delta\sigma = \sigma(\lambda_{ON}) - \sigma(\lambda_{OFF})$) in the DIAL equation (Fhjii and Fukuchi 2005)

$$N_x(R) = \frac{1}{2\Delta\sigma}\frac{d}{dR}\ln\frac{P(R,\lambda_{OFF})}{P(R,\lambda_{ON})}$$ (1)

Where $P(R,\lambda_{ON})$ and $P(R,\lambda_{OFF})$ the power backscattered signal received from distance R for both wavelengths. Special careful must be taken into account when selecting the adjacent wavelengths, where the different between the two wavelengths is preferred to be < 1 cm^{-1}, otherwise another two terms must be considered in the DIAL equation. DIAL, as a range resolved remote sensing technique, can detect lots of pollutants and greenhouse gases (H_2O, SO_2, O_3, CO, CO_2, NO, NO_2, CH_4, etc.) which play a big role in climate change and the earth's radiative budget. DIAL is possible in the UV (200 to 450 nm), the visible, and the near IR (1 to 5 micrometer), and in the mid-IR (5 to 11 micrometer). For example to measure Ozone as a green house gas with fatal direct effect on human health particularly in the troposphere, DIAL can be used in two appropriate bands; UV band (at 256 nm) and the mid-IR band (960 to 1070 cm^{-1}). DIAL operations advantages are successful both day and night, detecting gases and aerosol profiles simultaneously. It can be operated in ground, airborne, and space based platforms.

4.5 Doppler lidars

Atmospheric laser Doppler velocimetry including measurements of tornados, storms, wind, turbulence, global wind cycles, and the atmosphere temperature are some of the most important remote sensing techniques (Measures 1984). Doppler broadening is due to the

Doppler shift associated with the thermal motion of radiating (absorbing) species in the mesopause region such as Na, K, Li, Ca, and Fe (Measures 1984). Furthermore, the atmospheric temperature can be detected by measuring the Doppler broadening and the measured global wind pattern can be determined by measuring the Doppler shift of laser-induced florescence from atmospheric metals atoms such as Na in the middle and upper atmosphere (Bills 1991; She and Yu 1994). The use of Doppler broadening of the structure of Na D_2 line (by narrowband lidar) technique to determine the range resolved high resolution temperature profile of the mesopause region (75-115 km, is also called MLT for Mesosphere and Lower Thermosphere) and was proposed by Gibson et al. in 1979. The principle idea is that the absorption line will be broadened because of the Doppler effect for a single Na atom. Doppler broadened line is given by $\sigma_D = \sqrt{\dfrac{\kappa_B T}{M \lambda_0^2}}$, where M is the mass of a single Na atom, κ_B is the Boltzmann constant, λ_0 is the mean Na D_2 transition wavelength, and T is the temperature. As shown the Doppler broadened σ_D is a function of temperature. Therefore if we measure σ_D line-width, we can derive the temperature of the Na atoms in the mesopause which equal to the surrounding atmosphere temperature where Na atom is in equilibrium condition in the mesopause region (Fhjii and Fukuchi 2005).

4.6 Resonance fluorescence lidars

A Rayleigh lidar signal is useless above ~ 85 km, because of the low atmospheric density above that altitude. The backscattering cross section of Resonance fluorescence lidars is about 10^{14} times higher than Rayleigh backscattering cross-section for the same transmitter and receiver specifications, thus Resonance fluorescence lidars can be used in the upper atmosphere measurements. Resonance fluorescence lidars are measuring intensity at shifted wavelength using of Doppler technique (Bills 1991; She and Yu 1994) or Boltzmann technique (Gelbwachs 1994). Fluorescence lidar is used to measure metallic species in the upper layer of the atmosphere (~90km) such as, Na, K, Li (Jegou, M.Chanin et al. 1980), Ca and Fe (Granier, J. P. Jegou et al. 1989; Gardner, C. S. et al. 1993) and/or volcanic stratospheric aerosol, polar stratospheric clouds (PSCs), gravity waves, and stratospheric ozone layer. This lidar has high sensitivity and accuracy. It is also, used to determination of wind, temperature, and study of thermal structure and complex atmospheric dynamics.

5. Lidar wavelengths

Based on the wavelength that been used in lidar measurements, one can classify lidar into: Elastic, inelastic, multi wavelength, and femto-second white light lidars. Brief descriptions are introduced in the following sub-sections.

5.1 Elastic lidar

An elastic scattering is defined as light scattering with no apparent wavelength shift or change with the incident wavelength. Elastic backscatter lidar operation, as one of the most popular lidar systems, is based on the elastic scattering physical process. It is detecting the total atmospheric backscatter of molecular and particle together without separation. Hence, elastic backscattering lidar is the sum of Rayleigh and Mie scatterings. The main disadvantage of elastic lidar is the difficulty of separating Mie form Rayleigh signals. More details explain how to overcome this disadvantage are given as follow.

1. It is difficult to determine accurately the volume extinction coefficient of the particles or aerosol, where we can not separate Mie form Rayleigh signals. In this case, we have to assume the a value for particle lidar ratio, $S_a(R)$, where $S_a(R) = \dfrac{\alpha_a(R)}{\beta_a(R)}$, to solve the lidar equation for the aerosol extinction coefficient ($\alpha_a(R)$). This assumption is impossible to estimate reliably; since the aerosol lidar ratio $S_a(R)$ varies strongly with the altitude ($S_a(R)$ varies between 20 and 100) due to the relative humidity increment with the altitude (S ratio depends on chemical, physical and morphological properties of the particles which are relative humidity dependent). As shown in Table 1 (Kovalev and Eichinger 2004), big variations of aerosol typical lidar ratio for different aerosol types have been determine at 532 nm wavelength using Raman lidar. Figure 2 shows a lidar return signal on June 30, 2004 at the CCNY site. The figure also shows an example of the aerosol lidar ratio $S_a(R)$ retrieval between 20 and 100.

Aerosol (particle) types	Aerosol lidar ratio $S_a(R)$ (sr)
Marine Particle	20-35
Saharan dust	50-80
Less absorbing urban particles	35-70
Absorbing particles from biomass burning	70-100

Table 1. Different aerosol types and the corresponding aerosol lidar ratio $S_a(R)$

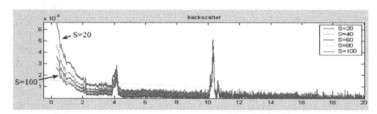

Fig. 2. CCNY lidar retrieval for $S_a(R)$ ration, June 30, 2004

To determine the aerosol lidar ratio $S_a(R)$, we can use (a) Raman lidar and High Spectra Resolution Lidar (HSRL) to get the extinction profile for particle then $S_a(R)$, Alternatively (b) sun-photometer observatory can be used to obtain the optical depth then seeking a solution by back integration. More details are given below.

a. Using Raman lidar and High Spectra Resolution Lidar (HSRL) to determine the extinction profile for particle and the particle backscatter coefficient can be obtained directly as well. These two lidars detect a separate backscatter signals from particle and molecular.
b. Using Sun-photometer observatory to obtain the optical depth (integration over the extinction coefficient profile) for both aerosol and molecule. Initially, in this method, we consider the reference boundary condition at the top of the lidar range is constant

(R_{max} , where the particle backscatter coefficient $\beta_a(R_{max})$ is negligible compared to the known molecular backscatter value). Second, we seek a solution by back integration (Klett 1981) that is more stable than the corresponding forward solution. Therefore, given the following data set $\{S_a, \beta_a(R_{max})\}$, the lidar signal can be inverted to obtain both $\beta_a(R), \alpha_a(R)$. Consequently, an estimation of the data set $\{S_a, \beta_a(R_{max})\}$ is required and the approach that used to analyze the lidar signals and estimate the optical coefficient error is outlined in (Hassebo et al, 2005). Finally, elastic scattering is unable to identify the gas species but can detect and measure particles and clouds (Fhjii and Fukuchi 2005).

5.2 Inelastic backscattering lidar

The transmitted wavelength is different than the detected wavelength on inelastic lidars. An example of inelastic lidar is Raman Lidar. A Raman signal is very weak; therefore Raman lidar operations are restricted to the nighttime due to the strong background solar radiations during the daytime. Three ways to overcome this difficulty, they are: (1) running Raman lidar within the solar-blind region (230-300 nm), (2) second is applying narrow-bandpass filter or Fabry-Perot interferometer, and (3) the third method is operating Raman lidar in the visible band of the spectra, during the daytime, and deduct the background solar radiation noise.

1. The first method is running Raman lidar within the solar-blind region (230-300 nm), where the ozone layer in the stratosphere (20-30 km) absorbs the lethal solar radiation in this spectral interval. Consequently, lidar can be operated diurnally in the solar-blind region without getting affected by the solar background noise. However, the main drawback of running lidar in this region is the attenuation of the transmitted and the returned signals by the stratospheric ozone. Another drawback is the eye hazard issue. Using this technique, in 1980th, there were some attempts to measure water vapor and temperature using multiwavelength in the solar-blind region (Renaut 1980; Petri 1982).
2. The second method is applying a narrow-bandpass filter or Fabry-Perot interferometer (Kovalev 2004). But the flitter will attenuate the signal strength as well. This is considered the main disadvantage of this method.
3. The third method has been proposed by Hassebo et al. in 2005 and 2006. The principal idea is to operate Raman lidar in the visible band (607 nm for N_2, 407 nm for water vapor, and 403 nm for liquid water vapor) of the spectra and then deduct the background solar radiation noise, simultaneously during the daytime optimally. This objective can be accomplished by using a polarization discrimination technique to discriminate between the sky background radiation noise and the backscattering signal. This can be approached using two polarizers at the transmitter and the receiver optics (Hassebo, B. Gross et al. 2005; Hassebo, Barry M. Gross et al. 2005; Hassebo, B. Gross et al. 2006). This technique improved the lidar Signal-to-Noise Ration (SNR) up to 300 %, and the attainable lidar range up to 34%. A discussion of this technique is introduced in section 2 of this chapter.

5.3 Multiple wavelength lidar

If the lidar transmitter is a single wavelength laser, the lidar is called single wavelength lidar. However, the lidar is referred to as a multiple wavelength lidar if it is transmitting more than one wavelength. All transmitted light into the atmosphere with wavelength shorter than 300 nm is absorbed by ozone and oxygen (solar-blind region). Wavelengths shorter than 300 nm are fatal wavelengths. Consequently the minimum wavelength for elastic lidar is approximately 300 nm. The commonly used wavelengths in lidar operations are near infrared (1064 nm), visible (532 nm), and ultraviolet (335 nm) for backscatter lidars, (607 nm) for N2, (407 nm) for water vapor, and (403 nm) for liquid water vapor. The multiple wavelength backscatter lidar can be used to distinguish between fine particles (emitted from fog, combustion, plume and burning smoke) and big particles such as water vapor or clouds. This differentiation can be achieved using angstrom coefficient (Hassebo, Y. Zhao et al. 2005).

Another example for multiple wavelength backscatter lidar is the DIfferential Absorption Lidar (DIAL). DIAL is used to measure concentrations of chemical species such as ozone, water vapor, and pollutants in the atmosphere. A DIAL lidar uses two distinct laser wavelengths which are selected so that one of the wavelengths is absorbed strongly by the molecule of interest while the other wavelength is not. The difference in intensity of the two return signals can be used to deduce the concentration of the molecule being investigated.

5.4 Femto-second white light lidar

Extremely high optical power (tera-watt) can be created from femto-second (1 fsec $=10^{-15}$ sec) laser pulse with 1mj energy. That is Femto-second white light lidar (fsec-lidar). In the era of global warming and climate change, fsec-lidar is used to detect and analyze aerosol size and aerosol phase (measuring the depolarization), water vapor, and for better understanding of forecasting, snow and rain. The inaccessibility to 3-D analysis is a disadvantage of Differential Optical Absorption Spectrometer (DOAS) and Fourier Transform Infrared spectroscopy (FTIR). This disadvantage has been conquered by Fsec-lidar white light lidar. At the same time it has the multi-component analysis capability of DOAS and FTIR by using a wide band light spectrum (from UV to IR); e.g. visible (Wöste, Wedekind et al. 1997; Rodriguez, R. Sauerbrey et al. 2002). An example of fsec-lidar, based on the well-known chirp pulse amplification (CPA) technique, is the 350 mJ pulse with 70 fsec duration and peak power of 5 TW at wavelength of 800 nm (Fhjii and Fukuchi 2005).

6. Purposes of lidar measurements

The purpose of Lidar Measurements is an additional way to classify lidars. Aerosol, clouds, and Velocity and Wind Lidars are introduced briefly in the following sub-sections.

6.1 Aerosol lidar

The atmosphere contains not only molecules but also particulates and aerosols including clouds, fog, haze, plumes, ice crystals, and dust. The aerosol is varied in radius; from a few nanometers to several micrometers. The bigger the aerosol size the more complex the calculations of their scattering properties. Aerosol concentration varies considerably with time, type, height, and location (Stephens 1994). Aerosols absorb and scatter solar radiation (all aerosols show such degree of absorption of the ultraviolet and visible bands) and provide cloud condensation sites (Charleon 1995). Aerosol absorption degree indicates the aerosol type. Atmospheric aerosol altitude, size, distribution and transportation are major global uncertainties due to their effects on controlling the earth's planet climate stability and global warming issues. In addition of the impact of aerosol in the atmospheric global climate change (Charlson, J. Langner et al. 1991; Charlson, S. E. Schwartz et al. 1992), it also affects human health with diseases such as lung cancer, bronchitis, and asthma. These have been essential motivations to study aerosol properties and transportation. Lidars have been successfully applied to study stratospheric aerosols mainly sulfuric-acid/water droplet (Zuev V., V. Burlakov et al. 1998), tropospheric mixture aerosols of natural (interplanetary dust particle and marine) and of anthropogenic (sulfate and soot particles) (Barnaba F and Gobbi 2001) and climate gases such as stratospheric ozone (Douglass L. R., M.R. Schoeberl et al. 2000) as well as for analyzing the clouds properties (Stein, Wedekind et al. 1999). Aerosol sources can origin from nitrate particles, sea-salt particles, and volcanic ashes and rubble. Aerosol particle sizes were categorized as aitken, large, and giant particles (Junge 1955), where: (a) Dry radii < 0.1 μm, Aitken particles, (b) Dry radii 0.1 μm< r < 1 μm, large particles, and (c) Dry radii r > 1 μm, giant particles. Aerosol concentration decreases with increasing altitude. 80% of the aerosols condense in the lowest two kilometers of the troposphere (i.e., within the Planetary Boundary Layer (PBL)) as shown in Fig 3, for New York City on August 11, 2005.

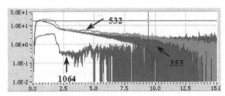 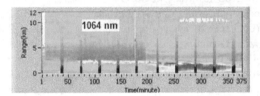

Source: CCNY lidar system

Fig. 3. New York City aerosol PBL, Aug 11, 2005

The extinction profile is considered a high-quality indicator (in the cloud free case) of aerosol concentration. A principle of measuring aerosol is using the wavelength between 300-1100 nm to determine the particle extinction and backscattering profiles. A good example for lidar, that has been used to monitor aerosol without attendance, is Micro-Pulse Lidar (MPL) (Spinhirne 1991; Spinhirne 1993; Welton, Campble et al. 2001). CUNY MPL at LaGuardia Community College will play a significant role in studying the impact of the anthropogenic aerosol on human health, life, air quality, climate change, and earth's radiation budget once it is deployed. High Spectral Resolution Lidar (HSRL) can be used, as well, to measure aerosol scattering cross section, optical depth, and backscatter phase function in the atmosphere. This can be achieved by separating the Doppler-broadened molecular backscatter return from the un-broadened aerosol return. The molecular signal is then used as a calibration target which is available at each point in the lidar profile.

6.2 Cloud lidar

Cloud particle radius is larger than 1 μm (between about 2 μm to around 30 μm), which is bigger than the lidar wavelength (300- 1100 nm). Therefore lidars cannot measure the cloud size distribution (Fhjii and Fukuchi 2005). However, lidars can detect the cloud ceiling, thickness, and its vertical profile where the lidar return signal from the cloud is very strong (because cloud behaves as obstruction in the laser propagation path).

As shown in Fig 4, using three wavelengths of 355, 532, and 1064 nm, the CCNY stationary lidar detected clouds vertical structure between 3.5 to 4.5 km height, and the planetary boundary layer (Welton, Campble et al.) on January 25, 2006.

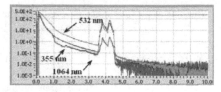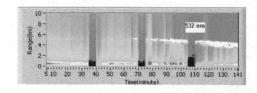

Source: CCNY lidar

Fig. 4. CCNY lidar data shows cloud ceiling, thickness, and structure, Jan 25, 2006

Clouds cover approximately 50% of the earth (Liou 2002). Based on the altitude (i.e., temperature) clouds are formed in liquid or solid (crystal) phases. Clouds and their interaction with aerosol and their impact on local and global climate change encouraged NASA to create various projects to monitor and study clouds distribution, thickness, transportation, and observe transitional form of clouds or combination of several forms and varieties. Cloud-Aerosol Lidar and Infrared Pathfinder Satellite Observations (CALIPSO), Micro-Pulse Lidar (MPL), and Polarization Diversity Lidar (PDL is a lidar with two channels to detect two polarizations (Fhjii, 2005; Sassen, 1994)) are well-known lidars to measure and detect clouds. Measuring cloud phase is based on Mie scattering theory, the backscattering from non-spherical (e.g., crystal phase) particles changes the polarization strongly, but the spherical (water droplets) particles do not (Sassen, K. et al. 1992; Sassen 1994). Both spherical and non-spherical cloud particles have a degree of depolarization ($\delta = I_\perp / I_{||}$) due to the multiple scattering effects, where $I_\perp, I_{||}$ are respectively the perpendicular and the parallel intensity components for the incident light. But non-spherical cloud particles degree of depolarization is greater than spherical particles depolarization ($\delta_{NS} \rangle \delta_S$). Polarization lidars are used to differentiate between cloud liquid and sold phases.

Fig 5, shows thin cloud signals that were provided by Hassebo et al. on January 10, 2006 using elastic Mie scattering stationary lidar at the City College of NY site (longitude 73.94 W, latitude 40.83 N), at 355, 532, and 1064 nm wavelengths. Comparing the thin cloud signal (Fig 5) with Fig 4 (thick cloud signal) we noted that, as a result of laser rapid attenuation while it is penetrating the cloud, in the thin cloud case the visible beam had a sufficient intensity to open a channel with high optical transparency to a higher altitude. In contrast, in Fig 4 the cloud was thick enough to prevent the laser beams from increasing their depth of penetration into the layer beyond the cloud. That explains the useless noisy (UV and IR) signals after the cloud ceiling (R= 4.5 km, and R= 11.5 km) in both cases, and for visible signal in the heavy cloud case even when the altitude is low (4.5 km). We noted also the PBL was shown clearly in both cases where the aerosol loading in New York City is always high.

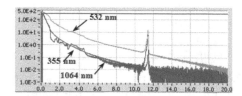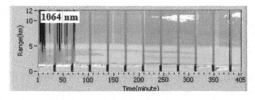

Fig. 5. CCNY Lidar backscattering signals show thin cloud at 11km, Jan 10, 2006.

6.3 Velocity and wind lidar

Doppler lidar can be used to provide the velocity of a target. When the light transmitted from the lidar hits a target moving towards or away from the lidar, the wavelength of the light reflected/scattered off the target will be changed slightly. This is known as a Doppler shift, hence Doppler lidar. If the target is moving away from the lidar, the returned beam will have a longer wavelength (sometimes referred to as a red shift). In other hand, if the target is moving towards the lidar the return light will be at a shorter wavelength (blue shifted). The target can be either a hard target or an atmospheric target. Thus the same idea is used to measure the wind velocity where, the atmosphere contains many microscopic dust and aerosol particles (atmospheric target) which are carried by the wind.

7. Lidar types based on platform

7.1 Ground-based lidar

The PBL is the most important layer to study in the earth's atmosphere. Ground-based lidar (stationary in laboratories and mobiles in vehicles stations) is providing us with continuous, stable, and high resolution measurements of almost most of the lower atmosphere parameters. Ground-based lidar has made an important contribution in correcting the satellite data and complete the missing parts from the satellite images. A good example in chapter 7 of this thesis shows how the ground-based lidar signature is supporting the satellite operations to discriminate between cloud (big particle) and smoke plume (fine particle) and to determine the plumes height and thickness which the satellite cannot provide. The main drawback of ground-based lidar is the limitation of running during the bad weather (rain or snow) and the air control regulations issues.

7.2 Air-borne lidar

Due to the uncertainty in validation of some remote sensing methodologies, particularly to detect cloud and measure its properties, from ground-based lidar stations, the in situ probes are useful techniques. Also the inaccessibility of the object of interest from the ground-based or space-based systems is the other reason to use air-borne lidars. The air-borne lidar platforms are air-craft, balloon, and helicopter. Applications of using air-borne lidars are to measure aerosol, clouds, temperature profile, metals in the mesopause, ozone in the stratosphere, wind, PSCs, H_2O, and on land, water depth, submarine track, oil slicks, etc (Fhjii and Fukuchi 2005). One of the disadvantages of these platforms is the vibration problem.

7.3 Space-based lidar

The ground–based lidar provides a one spot at a unique moment measurement on the earth surface. The air-borne lidars are limited in one country or specific region as well as restricted by the weather or some times politic circumstances. The merit of the space-based lidar is to give global and/or continental images of the earth's atmosphere properties, structure, and activities. Certainly, space-based lidar needs very sophisticated, extremely expensive equipment, especially for remotely control the unattended operations and adaptive optics issue. In additional to the extremely important understanding of global scale phenomena (H_2O and carbon cycles, climate change, global warming, etc.) we have gained, we can reach an inaccessible areas by air-borne and/or ground-based stations such as oceans, north and south poles.

8. Lidar configurations

Essentially, there are two basic configurations for lidar systems; monostatic and bistatic configurations.

8.1 Monostatic lidar

Monostatic configuration is the typical configuration for modern systems. It is employed with pulsed laser source providing very good vertical resolution and beam collimation compared with the bistatic configuration. In monostatic configurations, the transmitter and receiver are at the same location, (see Fig. 6). Monostatic systems can be classified into two categories, coaxial systems and biaxial systems. Monostatic system was first used in 1938.

8.1.1 Monostatic coaxial lidar

In the Monostatic coaxial configuration the axis of transmitter laser beam is coincident with the receiver's telescope Field Of View (FOV) as shown in Figure 6 (a). The main disadvantages in the Configuration are the detector saturation problem that occurs once the lidar laser beam is shot, the unwanted signal that is detected from reflection of the transmitted light at the transmitter optics in the top of the receiver telescope, and the portion of the images - for short range - that are blocked by the secondary mirror.

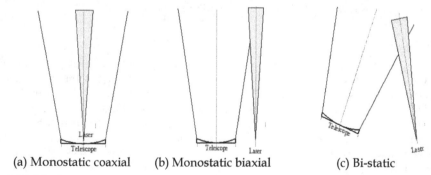

(a) Monostatic coaxial   (b) Monostatic biaxial   (c) Bi-static

Fig. 6. Field of view arrangements for lidar laser beam and detector optics

8.1.2 Monostatic biaxial lidar

In the Monostatic biaxial arrangement the transmitter and receiver are located adjacent to each other. Under this circumstance the laser beam will intersect with the receiver telescope VOF beyond specific range R. This range can be predetermined based on the distance between the laser FOV and telescope FOV axes. In fact, this configuration is quite useful in preventing the receiver photomultiplier (PMT) detectors saturation from the near-filed laser radiations (coaxial lidar disadvantage). However, in a biaxial lidar system, the detected signals are negatively affected by the geometrical form factor (GF) at shorter range. This effect makes near field measurements impossible (Measures 1984). Hassebo et al. proposed two techniques to overcome the problems of geometrical form factor (Hassebo, R. Agishev et al. 2004).

8.2 Bistatic configuration

Bistatic lidar configuration is involving a considerable separation between the laser transmitter and the receiver subsystems. However, the usefulness of this configuration was originally used in supporting lidar with continuous wave (cw) laser source to overcome the prevention of measuring of the height variation of the density caused by cw laser (Fhjii and Fukuchi 2005). Currently, this arrangement is rarely used (Measures 1984).

As a summary of some lidar physical processes, their corresponding applications and objective of measurements are given in Table 2.

Lidar Type Based On					
Process	Wavelength	Objective	Platform	Configurations	Other
Rayleigh Doppler	Elastic	Wind	Ground-based	Monostatic	Stratosphere Mesopause
Backscatter	inelastic	Cloud	Air-borne	Monostatic	Troposphere
Mie Backscatter Raman	single WL Multiple WL	Aerosol H_2O	Ground-based Space-based	Monostatic (Biaxial and Coaxial)	Troposphere Stratosphere
Raman	DIAL Raman DIAL	Ozone Humidity gaseous	Ground-based	Monostatic	Troposphere
Fluorescence		Wind /Heat flux	Air-borne	Monostatic	mesosphere

Table 2. Lidar classification and related research

9. Improve lidar signal-to-noise ratio during daytime operations

In this section, the impact and potential of a polarization selection technique to reduce sky background signal for linearly polarized monostatic elastic backscatter lidar measurements are examined. Taking advantage of naturally occurring polarization properties in scattered sky light, a polarization discrimination technique was devised. In this technique, both lidar transmitter and receiver track and minimize detected sky background noise while maintaining maximum lidar signal throughput. Experimental Lidar elastic backscatter measurements, carried out continuously during daylight hours at 532 nm, show as much as a factor of $\sqrt{10}$ improvement in signal-to-noise ratio (SNR) and the attainable lidar range up to 34% over conventional un-polarized schemes. Results show, for vertically pointing lidars, the largest improvements are limited to the early morning and late afternoon hours. The resulting diurnal variations in SNR improvement sometimes show asymmetry with solar angle, which analysis indicates can be attributed to changes in observed relative humidity that modifies the underlying aerosol microphysics and observed optical depth.

9.1 Introduction

This work describes a technique which is designed to improve the operation of conventional elastic backscatter lidars in which the transmitted signal is generally linearly polarized. The technique requires the use of a polarization sensitive receiver. Polarization selective lidar systems have, in the past, been used primarily for separating and analyzing polarization of lidar returns, for a variety of purposes, including examination of multiple scattering effects and for differentiating between different atmospheric scatterers and aerosols (Schotland, K. Sassen et al. 1971; Hansen and Travis 1974; Sassen 1974; Platt 1977; Sassen 1979; Platt 1981; Kokkinos and Ahmed 1989; G.P.Gobbi 1998; Roy, G. Roy et al. 2004). In the approach described here, the polarized nature of the sky background light is used to devise a polarization selective scheme to reduce the sky background power detected in a lidar. This leads to improved signal-to-noise ratios (SNR) and attainable lidar ranges, which are important considerations in daylight lidar operation (Hassebo, B. Gross et al. 2005; Hassebo, Barry M. Gross et al. 2005; Ahmed, Y. Hassebo et al. 2006; Ahmed, Yasser Y. Hassebo et al. 2006; Hassebo, B. Gross et al. 2006). The approach, discussed here, is based on the fact that most of the energy in linearly polarized elastically backscattered lidar signals retains the transmitted polarization (Schotland, K. Sassen et al. 1971; Hansen and Travis 1974; Kokkinos and Ahmed 1989), while the received sky background power (Welton, Campble et al.) observed by the lidar receiver shows polarization characteristics that depend on both the scattering angle, θ_{sc}, between the direction of the lidar and the direct sunlight and the orientation of the detector polarization relative to the scattering plane. In particular, the sky background signal is minimized in the plane perpendicular to the scattering plane, while the difference between the in-plane component and the perpendicular components (i.e degree of polarization) depends solely on the scattering angle. For a vertically pointing lidar, the scattering angle θ_{sc} is the same as solar zenith angle θ_s Fig. 7. The degree of polarization of sky background signal observed by the lidar is largest for solar zenith angles near $\theta_S \approx 90^o$ and smallest at solar noon. The essence of the proposed approach is therefore, at any time, to first determine the parallel component of the received sky background (Pb) with a polarizing analyzer on the receiver, thus minimizing the detected Pb, and then orienting the polarization of the outgoing lidar signal so that the polarization of the received lidar backscatter signal is aligned with the receiver polarizing analyzer. This ensures unhindered passage of the primary lidar backscatter returns, while at the same time minimizing the received sky background Pb, and thus maximizing both SNR and attainable lidar ranges.

The experimental approach and system geometry to implement the polarization discrimination scheme are described in the next Section. Section 1.8.3 presents results of elastic lidar backscatter measurements for a vertically pointing lidar at 532 nm taken on a clear day in the New York City urban atmosphere, that examine the range of application of the technique. In particular, the diurnal variations in Pb as functions of different solar angles are given and the SNR improvement is shown to be consistent with the results predicted from the measured degree of linear polarization, with maximum improvement restricted to the early morning and late afternoon. Section 1.8.4 examines the situations in which asymmetric diurnal variations in sky Pb are observed, and demonstrates the possibility that an increase in relative humidity (Halldorsson and Langerhoic), consistent with measured increases in measured Precipitable water vapor (PWV) and aerosol optical depth (AOD), may account for the asymmetry. Analysis of the overall results is presented in Section 1.8.5,

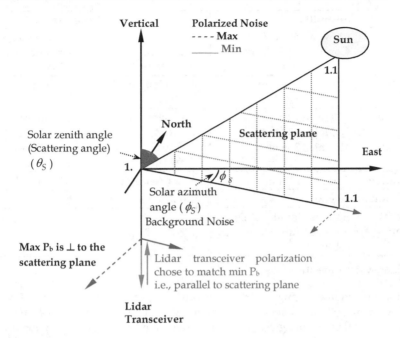

Fig. 7. Sky background suppression geometry for a vertical pointing lidar: θ_s is the solar zenith angle (equal to the scattering angle for this geometry) ϕ_s is the solar azimuth angle; and OAB is the solar scattering plane

where the SNR improvement factor is compared with a single scattering radiative transfer theory. Possible modifications due to multiple scattering are also explored.

In Section 1.8.6, the diurnal variation of the polarization rotation angle is compared to the theoretical result and an approach for automation of the technique based on theory is discussed. Conclusions and summary are presented in Section 1.8.7.

9.2 Experimental approach and system geometry

The City University of New York (CUNY) has developed two ground-based lidar systems, one mobile and one stationary, that operate at multiple wavelengths for monostatic elastic backscatter retrievals of aerosol and cloud characteristics and profiles. Lidar measurements are performed at the Remote Sensing Laboratory of the City College of New York, (CCNY). The lidar systems are designed to monitor enhanced aerosol events as they traverse the eastern coast of the United States, and form part of NOAA's Cooperative Remote Sensing Center (NOAA-CREST) Regional East Atmospheric Lidar Mesonet (REALM) lidar network. The lidar measurements, reported here, were carried out with the mobile elastic monostatic biaxial backscatter lidar system at the CCNY site (longitude 73.94 W, latitude 40.83 N), at 532 nm wavelength. The lidar transmitter and the receiver subsystems are detailed in Table 3.

The lidar return from the receiver telescope is detected by a photo-multiplier (PMT R11527P) with a 1 nm bandwidth optical filter (532F02-25 Andover), centered at the 532 nm

Transmitter		Receiver	
Laser	Q-Switched Nd: YAG Continuum Surelite ll-10	Telescope Aperture	CM_1400 Schmidt Cassegrian telescope 35.56 mm
Wavelength	1064, 532, 355 nm	Focal length	3910 mm
Energy/pulse	650 mj at 1064 nm 300 mj at 532 nm 100 mj at 355 nm	Detectors 532 nm 355 nm 1064 nm	Hamamatsu PMT: R11527 P PMT: R758-10 APD
Pulse Duration	7 ns at 1064 nm	Data Acquisition	LICEL TR 40-160
Repetition Rate	10 Hz	Photon Counting	LICEL TR 40-160
Harmonic Generation	Surelite Double (SLD) Surelite Third Harmonic (SLF)		

Table 3. Lidar system specifications

wavelength. For extended ranges, data is acquired in the photon counting (PC) mode, typically averaging 600 pulses over a one minute interval and using a Licel 40-160 transient recorder with 40 MHz sampling rate for A/D conversion and a 250 MHz photon counting sampling interval. Fig. 8 shows the arrangement used to implement the polarization-

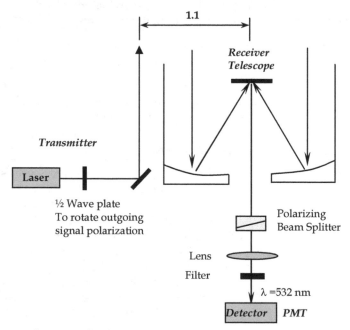

Fig. 8. Schematic diagram of polarization experiment set up for elastic biaxial monostatic lidar (mobile lidar system)

tracking scheme. To select the polarization of light entering the detector, a polarizing beam splitter is located in front of the collimating lens that is used in conjunction with a narrow band filter (alternatively, dichroic material polarizers were also used).

This polarizing beam splitter (analyzer) is then rotated to minimize the detected sky background Pb. Cross polarized extinction ratios on the receiver analyzer were approximately 10-4 . On the transmission side, a half wave plate at the output of the polarized laser output is then used to rotate the polarization of the outgoing lidar beam so as to align the polarization of the backscattered lidar signal with the receiver polarizing analyzer and hence maximize its throughput (i.e., at the minimum Pb setting). This procedure was repeated for all measurements, with appropriate adjustments being made in receiver polarization analyzer alignment and a corresponding tracking alignment in the transmitted beam polarizations to adjust for different solar angles at different times of the day, and hence minimize the detected Pb and maximize lidar SNR.

9.3 Results

Figures 9- to- 11 show experimental results with the receiver analyzer oriented to minimize Pb and a corresponding tracking lidar polarization orientation to maximize the detected backscattered lidar signal and its SNR at different times on Oct 07, 2004 (6:29 PM, 3 PM, and noon). All times given are in (EST) Eastern Standard Time.

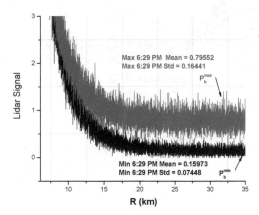

Fig. 9. Comparison of max Pb verses min Pb lidar signals at 6:29 PM on 07 October 2004.

The detected lidar signal is the sum of atmospheric backscatter of the laser pulse and the detected background light. The upper trace corresponds to the receiver polarization analyzer oriented to minimize Pb and the lidar transmitter polarization oriented to maximize the detected backscattered lidar signal while the lower trace is the result when orthogonal orientations of both receiver analyzer and lidar polarization are used, minimizing the sky background component in the return signal. Similar measurements were made at 3:00 PM and noon on the same day as shown in Figures 4 and 5 respectively.

Fig. 12 shows the resulting return signals in the far zone where the sky background signal is the dominant component (20-30 km range) for these times and for both orthogonal polarizations.

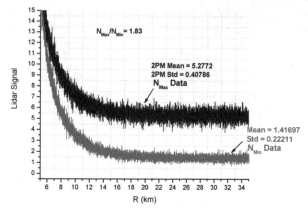

Fig. 10. Comparison of max max Pb (NMax) verses min Pb (NMin) lidar signals at 3 PM (EST) on 07 Oct 2004: Range 35 km, Lidar signal in linear scale

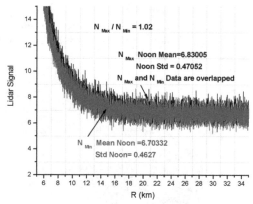

Fig. 11. Comparison of max Pb (NMax) verses min Pb (NMin) lidar signals at noon (EST) on 07 Oct 2004: Range 35 km, Lidar signal in linear scale, two signals are overlapped

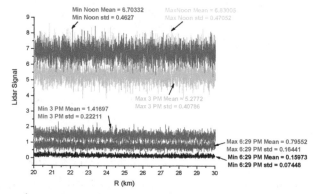

Fig. 12. Comparison of experimental return signals at 6:29 PM, 3 PM and noon on 07 Oct 2004, range of 20- 30 km, both orthogonal cases are shown.

The relative impact on the sky background signal, Pb, of the polarization discrimination scheme is seen to be largest at 6:29 PM, when the lidar solar angle is large (89°), while at noon it is minimal. The detected signal for maximum Pb is much noisier than the detected signal with minimum Pb, except in the noon measurement. This is consistent with the shot noise limit applicable to PMT's where the detected noise amplitude ΔP (standard deviation) is proportional to the square root of the mean detected background signal $\langle P \rangle$ (i.e., $\Delta P \propto \sqrt{\langle P \rangle}$) where P is the detector output, whose mean value is proportional to Pb. This relation is most conveniently expressed in terms of the ratios of the detected signals at the orthogonal polarization states $R = P_b^{max}/P_b^{min}$, in which the shot noise condition is now: $\Delta R = \sqrt{R}$. This relation has been verified in our experiments and the results summarized in Table 4.

Time	$\langle P_{min} \rangle$	ΔP_{min}	$\langle P_{max} \rangle$	ΔP_{min}	$R = \dfrac{\langle P_{max} \rangle}{\langle P_{min} \rangle}$	$\Delta R = \dfrac{\Delta P_{max}}{\Delta P_{min}}$	$\sqrt{R}$
Noon	6.7	0.46	6.83	0.46	1.2	1.019	1.09
3:00 PM	1.41	0.22	5.27	0.22	3.72	1.82	1.9
6:29 PM	0.159	0.074	0.795	0.074	5.2	2.2	2.2

Table 4. Comparison of experimental results to verify shot noise operation ($\Delta R = \sqrt{R}$)

In assessing the extent to which the polarization discrimination detection scheme can improve the SNR and the operating range, I compare the detected SNR with a polarizer, to that which would be obtained if no polarization filtering was used. When shot noise from background light is large compared to that from the lidar signal backscatter, the SNR improvement can be expressed in terms of an SNR improvement factor (G_{imp}) expressed in terms of maximum and minimum Pb measurements (P_b^{max}, P_b^{min}) as:

$$G_{imp} = \frac{SNR_{Max}}{SNR_{Unpol}} = \sqrt{\left(\frac{P_b^{min} + P_b^{max}}{P_b^{min}} \right)} = \sqrt{1 + \left(\frac{P_b^{max}}{P_b^{min}} \right)} \qquad (2)$$

To examine how the decreased Pb translates into a SNR improvement, Fig. 13 shows the range dependent SNR obtained for both maximum and minimum noise polarization orientations for a representative lidar measurement. The results show that for SNR=10, the range improvement resulting from polarization discrimination resulted in an increase in lidar operating range from 9.38 km to 12.5km (a 34% improvement). Alternatively, for a given lidar range, say 9 km, the SNR improvement was 250%.

Another useful way of looking at the effect of SNR improvement is to note that the SNR improves as the square root of the detector's averaging time. Thus a 250% improvement in SNR is equivalent to reducing the required averaging time by a factor of $(1 / 2.5)^2$.

9.4 SNR Improvement with respect to solar zenith angle

The SNR improvement factor (G_{imp}) is plotted as a function of the local time, Fig. 14, and the solar zenith angle, Fig.15. Since the solar zenith angle retraces itself as the sun passes

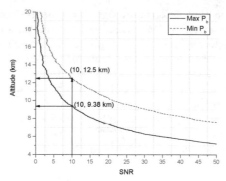

Fig. 13. Experimental range dependent SNR for maximum and minimum polarization orientations

through solar noon, it would be expected that the improvement factor (G_{imp}) would be symmetric before and after the solar noon and depend solely on the solar zenith angle. This symmetry is observed in Figs.14 and 15 for measurements made on 19 February 2005 and is supported by the relatively small changes in optical depth (AOD) values obtained from a collocated shadow band radiometer, (morning $\tau = 0.08$, afternoon $\tau = 0.11$)

Fig. 14. Gimp in detection wavelength of 532 nm verses local time on 19 February 2005

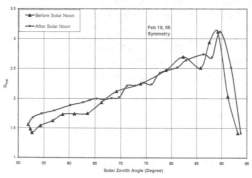

Fig. 15. Gimp in detection wavelength of 532 nm verses solar zenith angle on 19 February 2005

9.5 Effect of variable precipitable water vapor on SNR

Symmetry was, however, not always observed in our experimental results. Fig. 16 shows Gimp plotted as a function of the solar zenith angle for 23 February 2005. Small asymmetries were observed. These appear to be related to changes in humidity, which can modify the scattering properties and lead to enhanced multiple scattering effects. The results are supported by the variation in Precipitable water vapor (PWV) shown in Fig. 17, obtained from the CCNY Global Positioning System GPS measurements which were processed by the NOAA Forecast Systems Laboratory (FSL) (NOAA Web) for both days.

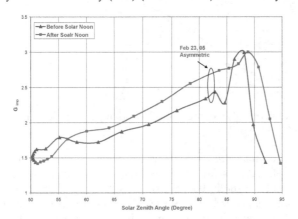

Fig. 16. Gimp in detection wavelength of 532 nm verses solar zenith angle on 23 February 2005

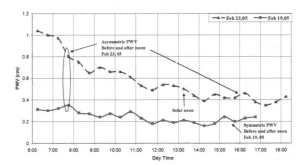

Fig. 17. PWV (cm) loading verses local time on 19 February 2005 and 23 February 2005

The 23 February the aerosol optical depth measurements from the shadow band radiometer show larger proportional changes (morning $\tau = 0.16$ afternoon $\tau = 0.09$) than those of 19 February, which are consistent with the asymmetry in the PWV, with higher optical depths corresponding to high PWV (and RH%) conditions.

9.6 SNR improvement azimuthally dependence

Within the single scattering theory, the polarization orientation at which the minimum Pb occurs should equal the azimuth angle of the sun (see Fig. 7). To validate this result, the polarizer rotation angle was tracked (by rotating the detector analyzer) over several seasons since February 2004 and compared with the azimuth angle calculated using the U.S. Naval Observatory standard solar position calculator (Applications) (14 April 2005). As expected, the polarizer rotation angle needed to achieve a minimum Pb closely tracks the azimuth angle, Fig. 18.

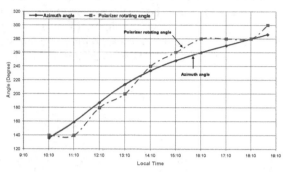

Fig. 18. Comparison between solar azimuth angle and angle of polarization rotation needed to achieve minimum Pb: 14 April 2005

This relationship is important since it allows us to conceive of an automated approach that makes use of a pre-calculated solar azimuth angle as a function of time and date to automatically rotate and set both the transmitted lidar polarization and the detector polarizer at the orientations needed to minimize Pb. With an appropriate control system, it would then be possible to track the minimum Pb by rotating the detector analyzer and the transmission polarizer simultaneously to maximize the SNR, achieving the same results as would be done manually as described above.

9.7 Conclusions and summary

SNR improvements can be obtained for lidar backscatter measurements, using a polarization selection/tracking scheme to reduce the sky background component. This approach can significantly increase the far range SNR as compared to un-polarized detection. This is equivalent to improvements in effective lidar range of over 30% for a SNR threshold of 10. The improvement is largest for large scattering angles, which for vertical pointing lidars occur near sunrise/sunset. Asymmetric skylight reduction sometimes observed in experimental results is explained by the measured increase in PWV and subsequent modification of aerosol optical depth by dehydration from morning to afternoon. It was also demonstrated that the orientation of the scattering plane defining the minimum noise state does not change in multiple scattering but follows the solar azimuth angle even for high aerosol loading. Therefore, it is quite conceivable to automate this procedure simply by using solar position calculators to orient the polarization axes.

10. Acknowledgment

I greatly would like to express my sincere appreciation and thankful to almighty God, Allah. Then, secondly, I am grateful to Drs. S Ahmed, B Gross, and Moshary, for their support during this research at The City University of New York. This work was supported under contract from NOAA # NA17AE1625.

11. References

Ahmed, S., Y. Hassebo, et al. (2006). *Examination of Reductions in Detected Skylight Background Signal Attainable in Elastic Backscatter Lidar Systems Using Polarization Selection.* 23rd International Laser Radar Conference (ILRC), Nara, Japan.

Ahmed, S. A., Yasser Y. Hassebo, et al. (2006). *Potential and range of application of elastic backscatter lidar systems using polarization selection to minimize detected skylight noise.* SPIE, Sweden.

Ansmann, A., U. Wanginger, M. Riebesell, C. Weitkamp amd W. Michaelis (1992). "Independent measurement of extinction and backscatter profiles in cirrus clouds by using a combined Raman elastic-backscatter lidar." *Appl. Opt.* 33: 7113-7131.

Applications, U. S. N. O. A. "U.S. Naval Observatory Astronomical Applications, http://aa.usno.navy.mil/data/docs/AltAz.html."

Barber, P., and C.Yeh (1975). "Scattering of Electromagnetic Waves by Arbitrarily Shaped Dielectric Bodies." *Appl. Opt* 14: 2864-2872.

Barnaba F and a. G. Gobbi (2001). "Lidar estimation of tropospheric aerosol extinction, surface area and volume: Maritime and desert-dust cases. ." *J. Geophys. Res.* 106 (D3): 3005-3018.

Bills, R., C. Gardner, and C. She (1991). "Narrowband lidar technique for sodium temprature and Doppler wind observations of the upper atmosphere." *Opt. Eng.* 30(a): 13-21.

Charleon, R. J. E. (1995). *Aeroeol forcing of climate.* New York, J. Wllay.

Charlson, R. J., J. Langner, et al. (1991). "Perturbation of the northern hemisphere radiative balance by backscattering from anthropogenic sulfate aerosols." *Tellus* 43AB: 152-163.

Charlson, R. J., S. E. Schwartz, et al. (1992). " Climate forcing by anthropogenic aerosols." *Science* 255: 423-430.

Deirmendjian, D. (1969). *Electromagnetic Scattering on Spherical Polydispersion.* New York.

Douglass I. R., M.R. Schoeberl, et al. (2000). "A composite view of ozone evolution in the 1995-1996 northern winter polar vortex developed from airborne lidar and satellite observations." *J. Geophys Res.* 106 (D9): 9879-9895.

Gobbi, G. P. (1998). "Polarization lidar returns from aerosols and thin clouds: a framework for the analysis." *Appl. Opt.* 37: 5505-5508.

Gardner, C. S., et al. (1993). " Simultaneous observations of sporadic E, Na, Fe, and Ca+ layers at Urbana, Illinois: Three case studies." *J. Geophys. Res.* 98: 16,865-16,873.

Gelbwachs, A. (1994). "Iron Boltzmann factor lidar: proposed new remote sensing technique for atmospheric temperature." *Appl. Opt*(33): 7151-7156.

Granier, C., J. P. Jegou, et al. (1989). "Iron atoms and metallic species in the Earth's upper atmosphere." *Geophys. Res. Lett.* 16: 243-246.

Halldorsson and a. J. Langerhoic (1978). "Geometrical form factors for the lidar function." *Appl. Opt* 17: 240-244.

Hamamatsu: http://www.hamamatsu.com

Hansen, J. and a. L. Travis (1974). "Light Scattering in Planetary Atmospheres." *Space Science Reviews* 16: 527-610

Hassebo, Y., R. Agishev, et al. (2004). *Optimization of Biaxial Raman Lidar receivers to the overlap factor effect"* Third NOAA CREST Symposium, Hampton, VA USA.

Hassebo, Y. Y., B. Gross, et al. (2005). *Polarization discrimination technique to maximize LIDAR signal-to-noise ratio.* Polarization Science and Remote Sensing II, SPIE

Hassebo, Y. Y., B. Gross, et al. (2006). "Polarization discrimination technique to maximize LIDAR signal-to-noise ratio for daylight operations." *App. Opt.* 45: 5521-5531.

Hassebo, Y. Y., Barry M. Gross, et al. (2005). *Impact on lidar system parameters of polarization selection / tracking scheme to reduce daylight noise.* Lidar Technologies, Techniques, and Measurements for Atmospheric Remote Sensing, SPIE.

Hassebo, Y. Y., Y. Zhao, et al. (2005). *Multi-wavelength Lidar Measurements at the City College of New York in Support of the NOAA-NEAQS and NASA-INTEX-NA Experiments* IEEE.

Heaps, W. S., J. Burris (1996). "Airborne Raman lidar." *Appl. Opt* 35: 7128-7137.

Heaps, W. S., J. Burris, and J. French (1997). "Lidar technique for remote measurement of temperature by use for a vibrational-rotational Raman spectroscopy." *Appl. Opt* 36: 9402-9405.

Jegou, J., M.Chanin, et al. (1980). "Lidar measurements of atmospheric lithium." *Geophys. Res. Lett.* 7: 995-998.

Jones, F. E. (1949). "Radar as an aid to the study of the atmosphere " *Royal Aeronautical Society* 53: 433-448.

Junge, C. (1955). "The size distribution and aging of natural aerosol as determined from electrical and optical data on the atmpsphere." *J. Meteorol* 12: 13-25.

Klett, J. D. (1981). "Stable analytical inversion solution for processing lidar returns." *Appl. Opt.* 20: 211–220.

Klett, J. D. (1985). "Lidar inversion with variable backscatter/extinction ratios" *Appl. Opt.* 24: 1638–1985.

Kokkinos, D. S. and S. A. Ahmed (1989). *Atmospheric depolarization of lidar backscatter signals.* Lasers '88' International Conference, Lake Tahoe, NV, STS Press.

Kovalev, V. and H. Moosmüller (1994). "Distortion of particulate extinction profiles measured with lidar in a two-component atmosphere." *Appl. Opt.* 33: 6499–6507.

Kovalev, V. and W. Eichinger (2004). *Elastic Lidar, Theory, Practice, and Analysis Mathods.* New Jersey, Wiley.

Liou, K. N. (2002). *An Introduction to Atmospheric Radiation.* California, Academic Press.

McClung, F. J. a. R. W. H. (1962). "Giant Optical Pulsations from Ruby." *Appl. Phys.* 33: 828-829.

Measures, R. M. (1984). *Laser Remote Sensing: Fundamentals and Applications.* NY, Wiley.

Measures. R. M., a. G. P. (1972). "A Study of Tunable Laser Techniques for Remote Mapping of Specific Gaseous Constituents of the Atmosphere." *Opto-electronics* 4: 141-153.

Middleton, W. E. K., and A.F.Spilhaus (1953). *Meteorological Instruments.* Toronto, , University of Toronto Press.

Mie, G. (1908). *Annalen der Physik* 24: 376-445.

MODIS Collection 5 Aerosol Retrieval Theoretical Basis Document.

NOAA-CREST " http://earth.engr.ccny.cuny.edu/noaa/wc/DailyData/."

NOAA " http://www.fsl.noaa.gov."

Petri, K., A. Salik, and J. Coony (1982). "Variable-Wavelength Solar-Blind Raman Lidar for Remote Measurement of Atmospheric Water-Vapor Concentration and Temprature." *Appl. Opt* 21: 1212-1218.

Platt, C. M. R. (1977). "Lidar observation of a mixed-phase altostratus cloud." *J. Appl. Meteorol.* 16: 339–345.

Platt, C. M. R. (1981). *Transmission and reflectivity of ice clouds by active probing.* Clouds, Their Formation, Optical Properties, and Effects, San Diego, Calif., Academic.

Renaut, J., C. Pourny, and R. Capitini (1980). "Daytime Raman-LidarMeasurements of Water Vapor." *Optics Letters* 5: 233-235.

Rodriguez, M., R. Sauerbrey, et al. (2002). "*Optics Letters.*" 27(772).

Roy, N., G. Roy, et al. (2004). "Measurement of the azimuthal dependence of cross-polarized lidar returns and its relation to optical depth." *Appl. Opt.* 43: 2777-2785.

Sassen, H. Z. K., et al. (1992). "Simulated polarization diversity lidar returns from water and precipitating mixed phase clouds." *Appl. Opt.* 31: 2914-2923.

Sassen, K. (1974). "Depolarization of laser light backscattered by artificial clouds." *J. Appl. Meterol.* 13: 923–933.

Sassen, K. (1979). "Scattering of polarized laser light by water droplet, mixed-phase and ice crystal clouds. 2. Angular depolarization and multiple scatter behavior." *J. Atmos. Sci* 36: 852-61.

Sassen, K. (1994). "Advanced in polarization diversity lidar for cloud remote sensing." *Proc. IEEE* 82: 1907-1914.

Sassen, K. and a. R. L. Petrilla (1986). "Lidar depolarization from multiple scattering in marine stratus clouds." *Appl. Opt.* 25: 1450– 1459.

Schotland, R. M. (1966). *Some Obsevation of the vertical Profile of Water Vapor by a Laser Optical Radar.* 4th Symposium on Remote Sensing of the Environment Univ. of Michigan.

Schotland, R. M., K. Sassen, et al. (1971). "Observations by lidar of linear depolarization ratios by hydrometeors." *J. Appl. Meteorol* 10: 1011–1017.

She, C. and a. J. Yu (1994). "Simultaneous three-frequency Na lidar measurements of radial wind and temperature in the mesopause region." *Geophys. Res. Lett.* 21: 1771-1774.

Spinhirne, J. D. (1991). *Lidar aerosol and cloud backscatter at 0.53, 1.06 and 1.54 μm.* presented at the 29th Aerospace Sciences Meeting, Reno, NV.

Spinhirne, J. D. (1993). "Micro pulse lidar." *IEEE TRANSACTIONS ON GEOSCIENCE AND REMOTE SENSING* 31: 48-54.

Stein, B., C. Wedekind, et al. (1999). "Optical classification, existence temperatures, and coexistence of different polar stratospheric cloud types." *J. Geophys. Res.* 104 (D19): 23983–23993.

Stephens, G. L. (1994). *Remote Sensing of the Lower Atmosphere: An Introduction.* New York, Oxford Univ. Press.

Takashi Fhjii and T. Fukuchi (2005). *Laser Remote Sensing,* Taylor and Francis Group

Velotta, R., B. Bartoli, et al. (1998). "Analysis of the receiver response in lidar measurements." *Appl. Opt.* 37: 6999-7007.

Welton, E., J. Campble, et al. (2001). First Annual Report: The Micro-pulse Lidar Woldwide Observational Network, Project Report

Wiscombe, W. J. (1980). "Improved Mie Scattering Algorithms." *Appl. Opt* 19: 1505.

Wöste, L., C. Wedekind, et al. (1997). "Laser und Optoelektronik " 29 (5)(51).
Zuev V., V. Burlakov, et al. (1998). "Ten Years (1986-1995) of lidar observations of temporal and vertical structure of stratospheric aerosol over Siberia." *J. Aerosol Sci.* 29 1179-1187.

3

Hyperspectral Remote Sensing – Using Low Flying Aircraft and Small Vessels in Coastal Littoral Areas

Charles R. Bostater, Jr., Gaelle Coppin and Florian Levaux

Marine Environmental Optics Laboratory and Remote Sensing Center,
College of Engineering, Florida Institute of Technology, Melbourne, Florida
USA

1. Introduction

Large field of view sensors as well as flight line tracks of hyperspectral reflectance signatures are useful for helping to help solve many land and water environmental management problems and issues. High spectral and spatial resolution sensing systems are useful for environmental monitoring and surveillance applications of land and water features, such as species discrimination, bottom top identification, and vegetative stress or vegetation dysfunction assessments[1]. In order to help provide information for environmental quality or environmental security issues, it is safe to say that there will never be one set of sensing systems to address all problems. Thus an optimal set of sensors and platforms need to be considered and then selected. The purpose of this paper is to describe a set of sensing systems that have been integrated and can be useful for land and water related assessments related to monitoring after an oil spill (specifically for weathered oil) and related recovery efforts. Recently collected selected imagery and data are presented from flights that utilize an aircraft with a suite of sensors and cameras. Platform integration, modifications and sensor mounting was achieved using designated engineering representatives (DER) analyses, and related FAA field approvals in order to satisfy safety needs and requirements.

2. Techniques

2.1 Imaging systems, sensor systems and calibration

Sensors utilized have been: (1) a photogrammetric 9 inch mapping camera utilizing a 12 inch focal length cone, and using AGFA X400PE1 color negative film that has been optimized for high resolution scanning (2400 dpi) in order to reduce the effects of newton rings and an associated special glass plate from Scanatronics in the Netherlands; (2) forward and aft full high definition (HD) video cameras recording to solid state memory with GPS encoding; (3) a forward mounted Nikon SLR 12.3 megapixel digital camera with a vibration reduction zoom lens and GPS encoding; (4) a high hyperspectral imaging system with 1376 spatial pixels and 64 to 1040 spectral bands.

The HSI imaging system utilizes a pen tablet computer with custom software. The HSI pushbroom system is integrated into the computer with an external PCMCIA controller card for operating the temperature stabilized monochrome camera which is bore sighted with a transmission spectrograph and ~39 degree field of view lens. The HSI imaging system is gimbal mounted and co-located with one of the HD 30 HZ cameras. The HSI system runs between ~20 to 90 HZ and is also co-located with a ~100 HZ inertial measurement unit (IMU). The IMU is strap down mounted to the HSI along the axis of view of the hyperspectral imager.

An additional 5HZ WAAS GPS output is recorded as another data stream into the custom software that allows on the fly changes to the integration time and spectral binning capability of the system. The HSI system is calibrated for radiance using calibration spheres and with spectral line sources for wavelength calibration. Flights are conducted with the 5 cameras in a fashion to allow simultaneous and or continuous operation with additional use of camera intervalometers that trigger the Nikon and photogrammetric camera. Examples of imagery taken on March 21, 2011 are shown below as well as spectral signatures and in-situ field targets that are typically utilized for processing imagery for subsurface or submerged water feature detection and enhancements.

Airborne imagery shown in this paper was collected at 1,225 m between 10 AM local time or 4 PM local, with a 1/225 second shutter speed and aperture adjusted for optimal contrast and exposure. The large format (9 in²) negatives scanned at 2400 dpi using a scanner and a special glass plate obtained from Scanatron, Netherlands allows for minimization of "newton rings" in the resulting ~255 megapixel multispectral imagery shown below (left image). Experience has shown that this method works well with AGFA X400PE1 film. The aerial negative scanning process is calibrated using a scanned target with known sub-millimeter scales 0.005 mm to 5 um resolution using a 2400 dpi scanner. The film scanning process results in three band multispectral images with spectral response curves published by the film manufacturer (Agfa or Kodak).

In-situ targets as shown in Figure 1 are used for calibration of the imagery using a combination of white, black or gray scale targets as shown below in an airborne digital image (right). Airborne targets are used for calibrating traditional film and digital sensor data for spatial and spectral characteristics using *in-situ* floating targets in the water as shown below.

Targets (figure 2) are placed along flight lines. These types of land and water targets are used for image enhancement techniques, for use as GPS georeferencing ground control points, and georeferencing accuracy assessments. They are necessary in order to assess shoreline erosion estimation resulting from oil spill impacts along littoral zones.

Figure 2 below shows images of weather oil taken in the Jimmy Bay area in January, 2011, eight months after the major spill was contained in the deep waters of the northern Gulf of Mexico.

2.2 Pushbroom imagery corrections for aerial platform motions

Airborne pushbroom imagery collected aboard moving platforms (ground, air, sea, space) requires geometric corrections due to platform motions. These motions are due to changes in the linear direction of the platform (flight direction changes), as well as sensor and platform motion due to yaw, pitch and roll motions. Unlike frame cameras that acquire a 2 dimensional image, pushbroom cameras acquire one scan line at a time. A sequence of

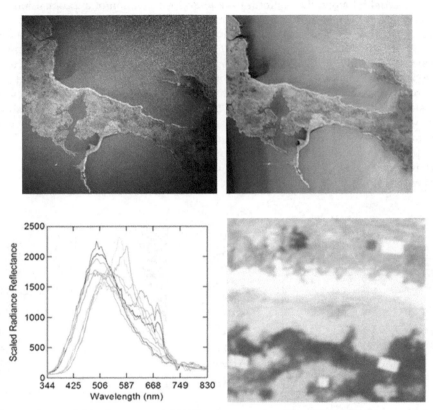

Fig. 1. Image (upper left) of a scanned AGFA color negative film (X400PE1) from an airborne flight March 21, 2011 over Barataria Bay, LA. The upper right image is a simultaneously collected hyperspectral RGB image (540, 532, 524 nm). Imagery indicates the ability to detect weathered oil in the area from oil spill remediation activities. The graph shows selected spectra in weathered oil impact areas. The lower right shows in-situ targets.

Fig. 2. Digital images of the *weathered oil* observed in early 2011 from the ground in the Barataria Bay, LA. areas shown above.

scan lines acquired along the platform track allows the creation of a 2 dimensional image. The consequence of using this type of imaging system is that the scan lines collected produce spatial artifacts due to platform motion changes - resulting in scan line feature offsets. The following describes the roll induced problem to be corrected. Consider an airplane that is flying over a straight road indicated by the dark red, vertical line in the left image below. Now assume the airplane or mobile platform undergoes unwanted platform roll motion and thus the resulting straight feature in the acquired scene is curved, as suggested by the light, blue line in the left image. One knows that the road was straight so the image as shown in Figure 3 (right) indicates a lateral scan line adjustment is required in order to straighten the feature (the blue line). This is accomplished by "shifting" the scan lines opposite to the platform roll motion and results in an image where the feature in the image is corrected. Thus, one needs to calculate the offset that corresponds to the shift the pixels undergo.

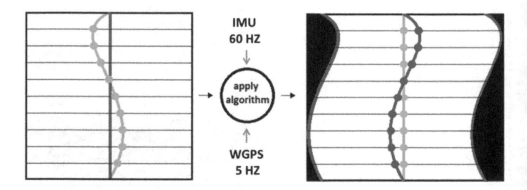

Fig. 3. The left figure shown in blue is a distorted road. The red line corresponds to the center of the scan line. The right image represents the corrected version of the left image. On this image the blue straight line is the road and the red curve is the actual position of the center pixels of the scan lines. In this example only the shift in the cross track direction is represented.

The offset mentioned previously can be corrected if sensing geometry and the hyperspectral imaging (HSI) system orientations are known when the different scan lines were taken. To obtain the platform and sensor orientation changes and position a 60 Hz update rate inertial measurement unit (IMU) was utilized and mounted to the gimbal mounted camera. An IMU is a device that is comprised of triads of accelerometers and gyroscopes. The accelerometers measure specific forces along their axes which are accelerations due to gravity plus dynamic accelerations. The gyroscopes measure angular rates. The IMU (Motion Node, GLI Interactive LLC, Seattle, Washington) that is used also has 3 magnetometers and outputs the orientation immediately by using those 3 types of sensors. In addition, differential WAAS 5 HZ GPS position, directional deviations, altitude with respect to a specified datum, and platform speed are collected during the flights.

An adaptive Kalman filter is used to estimate the induced platform motions using the combined sensor data from the GPS and IMU. The filtering technique thus allows one to obtain the relative position of each scan line and the corresponding spatial pixel shift that needs to be applied to correct the image. When a gimbal mounted HSI pushbroom camera is used, there are two main influences that cause the geometric distortions. These are the slowly varying directional changes of the platform and the roll induced motions. The first step in the algorithm is to use the GPS to calculate the position of the sensor (O_x, O_y, O_z) at every scan line. The second step accounts for the influence of the roll motion by using the IMU sensor data. The position of a pixel on the earth's surface can be estimated using:

$$x = Ox + \frac{s_x}{s_z}(h_{DEM} - Oz)$$

$$y = Oy + \frac{s_y}{s_z}(h_{DEM} - Oz) \tag{1}$$

$$z = h_{DEM}$$

Where (S_x, S_y, S_z) are components of a unit central scan line ray vector, (x, y, z) the position in meters compared to the origin (the initial position of the center of the scan line) and h_{DEM} is the surface elevation given in meters with respect to Mean Sea Level (MSL).

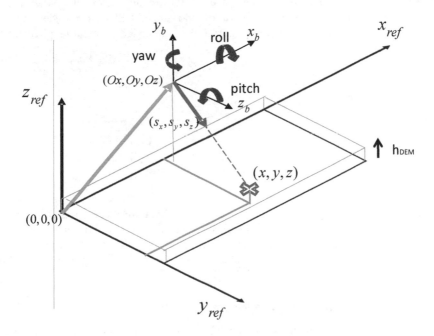

Fig. 4. This figure shows the position of the sensor (O_x, O_y, O_z) and the unit scan line ray vector in the reference coordinate system as well as the body (sensor platform) coordinate system with the possible platform motions. The position (x, y, z) is the position of the surface in the reference coordinate system that is located at the center of a HSI pixel .

The reference coordinate system chosen in this paper is a local tangent plane with the x axis pointed in the initial along track direction, y axis is 90 degrees clockwise to the x axis and corresponds to the initial cross track direction. In the results that are presented in this paper, shifts have only be applied in the cross-track direction. The shifts in meters are scaled to shifts in pixels as a function of the altitude (given by the GPS in meters), the field of view of the sensor (dependent upon the lens used) and the number of pixels in one scan line.

In the following section a description of the system is given, as well as the assumptions made. Then the application of the Kalman filter to acquire the position and velocity of the sensor is described with a detailed description of the vectors and matrices used. In the 2nd paragraph of this section, the influence of roll is taken into account. Then a paragraph that describes the image resampling phase applied to low flying airborne imagery in littoral areas.

In general, the application of the Kalman filter is used to acquire the position and velocity of the sensor is described with a detailed description of the vectors and matrices used and influence of roll is taken into account as described below, followed by a nearest neighbor resampling of the HSI imagery for each band independently.

Results that are presented in this paper, pixel shifts are only applied in the cross-track direction. Use of a gimbal sensor mount has allowed reduced HSI sensor motion corrections, however the need for improving image corrections in order to include pitch and yaw motions have been developed.

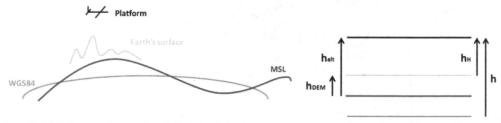

Fig. 5. This figure shows the different altitudes and heights used. Where h is the altitude of the platform with respect to the WGS84, h_{DEM} is the surface elevation, h_{alt} is the altitude of the platform with respect to MSL and h_H is the true altitude with respect to the earth's surface. In our applications we consider that the surface elevation is negligible as we take images of oil spills around MSL, so $h_H \approx h_{alt}$.

2.3 Description of the platform dynamic system

In order to model the movement of the platform, a discrete dynamic system described by the canonical state space equations are used:

$$\mathbf{x}_{k+1} = A_k \mathbf{x}_k + B_k \mathbf{u}_k + \mathbf{w}_k \qquad (2)$$

$$\mathbf{z}_k = H_k \mathbf{x}_k + \mathbf{v}_k \qquad (3)$$

where:

x_k = the state vector (6x1 matrix). $x_k = (O_x \ O_y \ O_z \ V_x \ V_y \ V_z)_k^T$ contains the position of the sensor (O_x, O_y, O_z) (in meters) and the velocity (V_x, V_y, V_z) (in meters per second) in the reference coordinate system.

A_k = the (6x6) matrix that gives the relation between the previous state vector to the current state vector when no noise and no input vector are considered . This relation can also be given below by equation (4) for x, y and z.

$$(O_x)_{k+1} = (O_x)_k + (V_x)_k \tag{4}$$

where:

Δt_k = the time-interval (in seconds) between step k and k+1.

B_k = the (6xm) matrix that relates the optional input vector u to the current state. (m = the number of elements in the control input vector if external forces are considered).

u_k = the control input vector (mx1 matrix), we assume that there are no external forces that act upon the system so u_k =0 in our application. Actually, it is assumed that the drag is exactly compensated by the thrust and gravity by the lift.

z_k = the measurement vector (6x1 matrix). $z_k = = (O_{xm} \ O_{ym} \ O_{zm} \ V_{xm} \ V_{ym} \ V_{zm})_k^T$ contains the position of the sensor (O_{xm}, O_{ym}, O_{zm}) (in meters) and velocity (V_{xm}, V_{ym}, V_{zm}) (in meters per second) in the reference coordinate system obtained by the GPS.

H_k = the measurement sensitivity (6x6) matrix also known as the observation matrix that relates the state vector to the measurement vector ($z_k = H_k x_k$).

w_k = the process noise or also called dynamic disturbance noise (6x1 matrix) which is assumed white and Gaussian with covariance matrix Q_k (6x6 matrix).

v_k = the measurement noise of the GPS (6x1 matrix) which is also assumed white and Gaussian (detailed calculations of the covariances see below) and its associated covariance matrix R_k (6x6 matrix).

The subscript k refers to the time step at which the vector or matrix is considered and indicates the time dependence.

x_k contains the real position and velocity, whereas z_k contains the measured position and velocity. z_k is thus always prone to measurement noise.

In the first step, GPS data is used to calculate the position of the sensor (O_x, O_y, O_z) the matrices are defined as follows in the reference coordinate frame:

$$A_k = \begin{pmatrix} 1 & 0 & 0 & \Delta t & 0 & 0 \\ 0 & 1 & 0 & 0 & \Delta t & 0 \\ 0 & 0 & 1 & 0 & 0 & \Delta t \\ 0 & 0 & 0 & 1 & 0 & 0 \\ 0 & 0 & 0 & 0 & 1 & 0 \\ 0 & 0 & 0 & 0 & 0 & 1 \end{pmatrix}_k, \quad Q_k = \begin{pmatrix} 0.1 & 0 & 0 & 0 & 0 & 0 \\ 0 & 0.1 & 0 & 0 & 0 & 0 \\ 0 & 0 & 0.1 & 0 & 0 & 0 \\ 0 & 0 & 0 & 0.1 & 0 & 0 \\ 0 & 0 & 0 & 0 & 0.1 & 0 \\ 0 & 0 & 0 & 0 & 0 & 0.1 \end{pmatrix}_k$$

Detailed calculations of the covariance's σ_v^2 and σ_h^2 are respectively the covariance in vertical and horizontal position (in m²) given by the GPS.

$$H_k = \begin{pmatrix} 1 & 0 & 0 & 0 & 0 & 0 \\ 0 & 1 & 0 & 0 & 0 & 0 \\ 0 & 0 & 1 & 0 & 0 & 0 \\ 0 & 0 & 0 & 1 & 0 & 0 \\ 0 & 0 & 0 & 0 & 1 & 0 \\ 0 & 0 & 0 & 0 & 0 & 1 \end{pmatrix}_k , \quad R_k = \begin{pmatrix} \frac{\sigma_h^2}{2} & 0 & 0 & 0 & 0 & 0 \\ 0 & \frac{\sigma_h^2}{2} & 0 & 0 & 0 & 0 \\ 0 & 0 & \sigma_v^2 & 0 & 0 & 0 \\ 0 & 0 & 0 & \frac{\sigma_{Vh}^2}{2} & 0 & 0 \\ 0 & 0 & 0 & 0 & \frac{\sigma_{Vh}^2}{2} & 0 \\ 0 & 0 & 0 & 0 & 0 & \sigma_{Vv}^2 \end{pmatrix}_k ,$$

The covariance's of the vertical and horizontal velocities σ_{Vv}^2 and σ_{Vh}^2 (in m² per seconds²) are however not given by the GPS but are calculated using using the following, where:

A given quantity y is a function of x_1, x_2, ... x_N given by the formula $y=f(x_1,x_2...,x_N)$. The uncertainties in x_i are respectively e_1, e_2, ..., e_N. The absolute uncertainty e_y is then given by

$$\left(e_y\right)^2 = \left(\frac{\partial f}{\partial x_1}\right)^2 \left(e_1\right)^2 + \left(\frac{\partial f}{\partial x_2}\right)^2 \left(e_2\right)^2 + ... + \left(\frac{\partial f}{\partial x_N}\right)^2 \left(e_N\right)^2$$

From the above, one has for the vertical, z direction $(V_z)_k = \frac{(O_z)_{k+1}-(O_z)_k}{\Delta t_k}$ and hence the covariance of the vertical velocity $(\sigma_{Vv}^2)_k$ equals $\frac{(\sigma_v^2)_{k+1}+(\sigma_v^2)_k}{\Delta t_k}$ since the velocity at time k equals the difference between the position at time k+1 and k, divided by the time interval. It is assumed that there is no uncertainty on the time interval. This is valid because one assumes that x_k and x_{k+1} are statistically independent. In an similar manner, one can calculate the covariance of the horizontal velocity, where one assumes

$$\sigma_{Ox}^2 = \sigma_{Oy}^2 = \frac{\sigma_h^2}{2} \text{ from } \sigma_h^2 = \sigma_{Ox}^2 + \sigma_{Oy}^2.$$

3. Kalman filter and smoothing approach

3.1 Position and velocity estimations

The Kalman filter consists of 2 steps. A temporal update step (also known as the "a priori" prediction step) and a measurement update step (also known as the "a posteriori" correction step). In the temporal step given by equations 4 and 5, the estimated state vector $\hat{x}_k^-$ and the estimation covariance P_k^- at time step k are predicted based on the current knowledge at time step k-1.

The state vector $\hat{x}_k$ contains the estimated position of the sensor (O_x, O_y, O_z) (in meters) and the velocity (V_x, V_y, V_z) (in meters per second) in the reference coordinate system.

The predictive procedure step is given by:

$$\hat{\mathbf{x}}_k^- = A_{k-1}\hat{\mathbf{x}}_{k-1}^+$$
$$P_k^- = A_{k-1}P_{k-1}^+ A_{k-1}' + Q_{k-1}$$

$$(5)$$

and the measurement update step (given by equation 6 below) corrects the predicted estimated $\hat{\mathbf{x}}_k^-$ and P_k^- using the additional GPS sensor measurements z_k to obtain the corrected estimate of the state vector $\hat{\mathbf{x}}_k^+$ and P_k^+ or:

$$K_k = P_k^- H_k' \left(H_k P_k^- H_k' + R_k \right)^{-1}$$
$$\hat{\mathbf{x}}_k^+ = \hat{\mathbf{x}}_k^- + K_k \left(\mathbf{z}_k - H_k \hat{\mathbf{x}}_k^- \right)$$
$$P_k^+ = \left(I - K_k H_k \right) P_k^-$$

$$(6)$$

where, $\hat{\mathbf{x}}_k^-$ and $\hat{\mathbf{x}}_k^+$ are respectively the predicted (-) and corrected (+) value of the estimated state vector (6x1 vector), P_k^- and P_k^+ (a 6 x 6 matrix) are respectively the predicted and corrected value of the estimation covariance of the state vector, or:

$$P_k^+ = diag\left(\sigma_{Ox}^2, \sigma_{Oy}^2, \sigma_{Oz}^2, \sigma_{Vx}^2, \sigma_{Vy}^2, \sigma_{Vz}^2 \right)_k$$

K_k (equation 6) is the Kalman gain (6x6 matrix).

The Kalman filter thus computes a weighted average of the predicted and the measured state vector by using the Kalman gain K_k. If one has an accurate GPS sensor, the uncertainty on the measurement will be small so there will be more weight given to the measurement and thus the corrected estimate will be close to the measurement. When one has a non-accurate sensor, the uncertainty on the measurement is large and more weight will be given to the predicted estimate.

A Kalman smoother has been applied as well where the equations are shown in (7) below. A Kalman Smoother in addition to the past observations also incorporates future observations to estimate the state vector:

$$C_k = P_k^+ A_k^T \left(P_{k+1}^- \right)^{-1}$$
$$\hat{\mathbf{x}}_k^s = \hat{\mathbf{x}}_k^+ + C_k \left(\hat{\mathbf{x}}_{k+1}^s - A_k \hat{\mathbf{x}}_k^+ \right)$$
$$P_k^s = P_k^+ + C_k \left(P_{k+1}^s - P_{k+1}^- \right) C_k^T$$

$$(7)$$

where:

$\hat{\mathbf{x}}_k^s$ = the smoothed estimated state vector (6x1 matrix).
P_k^s = the covariance (6x6) matrix of the smoothed estimated state vector.
C_k = the (6x6) matrix that determines the weight of the correction between the smoothed and non-smoothed state.

4. Roll correction

The second step in the algorithm for motion correction accounts for the influence of the roll motion by using the IMU orientation output. This is not included in the first Kalman filter because the IMU data is given at a higher frequency than the GPS data.

The state equations and Kalman filter/smoothing equations are given by 2.5, 2.6 and 2.7 with state vector $\hat{x}'_k$ containing the estimated position of the center pixel of the scanline on the surface (in meters) and the tangent of the roll angle (nondimensional) in the reference coordinate system. The measurement vector z_k' contains the position of the sensor (O_x, O_y) (in meters) specified by the output of the previous Kalman filter in the reference coordinate system and the tangent of the rollangle r_m (nondimensional) given by the orientation output of the IMU.

The matrices used are defined by:

$$
x'_k = \begin{pmatrix} x \\ y \\ \tan r \end{pmatrix}_k, \quad
A'_k = \begin{pmatrix} 1 & 0 & 0 \\ 0 & 1 & 0 \\ 0 & 0 & 1 \end{pmatrix}_k, \quad
Q'_k = \begin{pmatrix} 0.1 & 0 & 0 \\ 0 & 0.1 & 0 \\ 0 & 0 & 0.001 \end{pmatrix}_k
$$

$$
z'_k = \begin{pmatrix} O_x \\ O_y \\ \tan r_m \end{pmatrix}_k, \quad
H'_k = \begin{pmatrix} 1 & 0 & 0 \\ 0 & 1 & h_{alt} \\ 0 & 0 & 1 \end{pmatrix}_k, \quad
R'_k = \begin{pmatrix} \sigma^2_{Ox} & 0 & 0 \\ 0 & \sigma^2_{Oy} & 0 \\ 0 & 0 & \sigma^2_r \end{pmatrix}_k
$$

where:

r   = the roll angle (in radians).
h_{alt} = the altitude of the sensor (in meters) with respect to MSL.
σ^2_{Ox} and σ^2_{Oy} = respectively the covariance of the position in x and y direction of the sensor given by the previous Kalman filter (in m²).
σ^2_r = the covariance of the roll angle given by the IMU (nondimensional).

5. Image resampling

In some cases, it is only desirable to cross-track shift corrections and not resample the image in order to keep the pure spectral signatures of measured pixels. Otherwise, 2D nearest neighbourhood resampling is used.

The cross-track shift corrections (which are in the y direction) s_s on the surface in meters need to be converted to pixelshifts s_p. The number of pixels in one scanline n_N, the altitude of the sensor above the surface in meters h_H and half of the angular field of view α are used. This is accomplished by defining a conversion ratio c_r, the shift in meters on the surface of 1 pixel shift, or:

$$
c_r = \frac{w}{\dfrac{n_N}{2}} \tag{8}
$$

where:

$w = h_H \tan \alpha$.

The pixelshift s_p is then given by $s_p = \dfrac{s_s}{c_r}$ as depicted below:

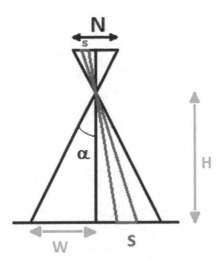

Fig. 6. This image shows the conversion triangles used to calculate the shiftratio between the shift on the earth surface s_s in meters and the pixelshift s_p. h_H is the altitude above the surface, α half of the angular field of view of the camera and n_N the number of pixels in one line.

6. Feature detection in hyperspectral images using optimal multiple wavelength contrast algorithms

Hyperspectral signatures and imagery offer unique benefits in detection of land and water features due to the information contained in reflectance signatures that directly show relative absorption and backscattering features of targets. The reflectance spectra that will be used in this paper were collected *in-situ* on May 31st 2011 using a SE590 high spectral resolution solid state spectrograph and the HSI imaging system described above. Bi-directional Reflectance Distribution Function (BRDF) signatures were collected of weathered oil, turbid water, grass and dead vegetation. The parameters describing the function in addition to the wavelength λ, (368-1115 nm) were the θ_i (solar zenith angle) = 71.5°, θ_0 (sensor zenith angle) = 55°, ϕ_i (solar azimuth angle) = 105° and the ϕ_0 (sensor azimuth angle) = 270°. The reflectance BRDF signature is calculated from the downwelling radiance using a calibrated Lambertian diffuse reflectance panel and the upwelling radiance at the above specified viewing geometry for each target (oil, water, grass, dead vegetation) as described in the figure below.

The figures below show the results of measurements from 400 to 900 nm for a 1 mm thick surface weathered oil film, diesel fuel, turbid water (showing the solar induced fluorescence line height feature, dead vegetation, and field grass with the red edge feature common to vegetation and associated leaf surfaces. These BRDF signatures are used below to select optimal spectral channels and regions using optimally selected contrast ratio algorithms in order to discriminate oil from other land & water features in hyperspectral imagery.

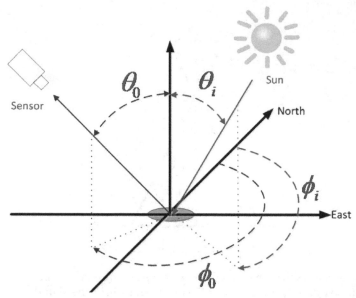

Fig. 7. Illumination and viewing geometry defined for calculation of the BRDF signatures collected using the 252 channel SE590 high spectral and radiometric sensitivity solid state spectrograph and the hyperspectral imaging system, where θ_i is the incident solar zenith angle of the sun, θ_0 is the sensor zenith angle, $\emptyset_i$ is the solar azimuth angle from the north and $\emptyset_0$ is the sensor azimuth angle as indicated above. In general, a goniometer measurement system is used to measure the BRDF in the field or laboratory environment as the sensor zenith and azimuth angles are changed during a collection period with a given solar zenith conditions.

The above BRDF signatures were used to select optimal spectral regions in order to apply the results to hyperspectral imagery collected from a weathered oil impacted shoreline in Barataria Bay, LA. The first method used was to perform feature detection using spectral contrast signature and HSI image contrast. The well know Weber's contrast definition is first used to determine the maximum (optimal) value of the contrast between a target t and a background b as a function of wavelength, or:

$$C_t(\lambda_k) = \frac{BRDF_t(\theta_0,\phi_0,\theta_i,\phi_i,\lambda_k) - BRDF_b(\theta_0,\phi_0,\theta_i,\phi_i,\lambda_k)}{BRDF_b(\theta_0,\phi_0,\theta_i,\phi_i,\lambda_k)} \tag{9}$$

The resulting contrast calculated across the spectrum for each channel are shown below using the 1 mm thick oil film as the target and the backgrounds of turbid water, dead vegetation (dead foliage), and field grass.

The result of the optimization of the contrast obtained from equation 9 yields an optimal channel and/or spectral region as a function of wavelength where the contrast is maximized between a specified target and specified background or feature in a hyperspectral image collected from a fixed or moving platform.

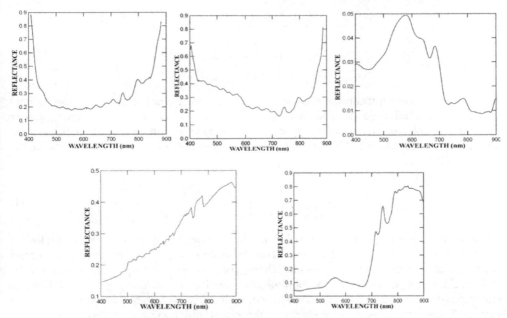

Fig. 8. Averaged (n=360) BRDF reflectance spectrums collected using a SE590 solid state spectrograph May 31, 2010. From upper left to right: BRDF spectrum of weathered oil (1 mm thick film) on clear water, diesel film (1mm thick film) on clear water, turbid water, with high chlorophyll content as indicated by the solar induced fluorescence line height, dead vegetation (dead leaves) and field grass showing the red edge. Solar angles were determined from DGPS location, time of day, and sensor position angles and measured angle from magnetic north direction.

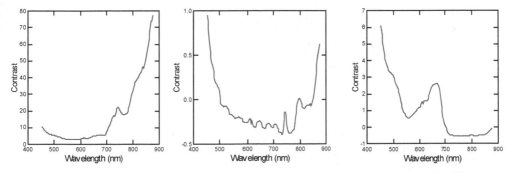

Fig. 9. Resulting BRDF Weber contrast signatures between oil as the target and different backgrounds (left to right): turbid water, dead vegetation (dead foliage) and field grass.

A limitation with this common definition of the contrast is that one band is used out of all the possible combinations available in a hyperspectral image for the feature detection or extraction algorithm. This limitation can be overcome, by defining an advantageous *"multiple-wavelength (or channel) contrast"* as:

$$C_t\left(\lambda_{k,m}\right) = \frac{BRDF_t\left(\theta_0,\phi_0,\theta_i,\phi_i,\lambda_k\right) - BRDF_b\left(\theta_0,\phi_0,\theta_i,\phi_i,\lambda_{k\pm m}\right)}{BRDF_b\left(\theta_0,\phi_0,\theta_i,\phi_i,\lambda_{k\pm m}\right)}$$

$$= \frac{BRDF_t\left(\theta_0,\phi_0,\theta_i,\phi_i,\lambda_k\right)}{BRDF_b\left(\theta_0,\phi_0,\theta_i,\phi_i,\lambda_{k\pm m}\right)} - 1 \tag{10}$$

The result of the optimization of this "*multiple-wavelength contrast algorithm*" is the optimal selection of a band ratio (located in a spectral region) minus one. Furthermore, a new definition of the inflection contrast spectrum (a numerical approximation of the second derivative) can be defined. The contrast inflection spectrum described in previous papers was given by:

$$I_t\left(\lambda_{k,m,n}\right) = \frac{BRDF_t\left(\theta_0,\phi_0,\theta_i,\phi_i,\lambda_k\right)^2}{BRDF_t\left(\theta_0,\phi_0,\theta_i,\phi_i,\lambda_{k\pm m}\right)BRDF_t\left(\theta_0,\phi_0,\theta_i,\phi_i,\lambda_{k-n}\right)} \tag{11}$$

where m and n are respectively defined as a dilating wavelet filter forward and backward operators described by Bostater, 2006. This inflection is used to estimate the second derivative of reflectance spectra. The underlying goal of computing an approximation of the second derivative is to utilize the nonlinear derivative based, dilating wavelet filter to enhance the variations in the reflectance spectra signals, as well as in the contrast spectrum signals. These variations directly represent the target and background absorption (hence: concave up) and backscattering (hence: concave down) features within a hyperspectral reflectance image or scene and form the scientific basis of the discrimination based noncontact optimal sensing algorithms. A practical limitation encountered using this definition above, is that a concave-down (or backscattering) feature value of the inflection as defined in 2.7 is greater than one and a concave up (or absorption) feature, in the inflection or derivative based wavelet filter defined in 1.7 will be between 0 and 1. There is thus a difference in scale between a concave-up and a concave-down behavior. Consider the following example:

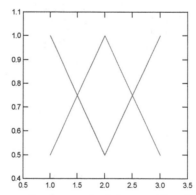

Fig. 10. Example concave-down (backscattering) feature (blue line) and a concave-up (absorption) feature (red line) of the same amplitude (Y axis) as a function of an spectral wavelength on the x axis.

In the case of the concave-down, the result of the inflection is:

$$I = \frac{1^2}{0.5 * 0.5} = 4$$

While in the case of the concave-up (same concavity), the result will be:

$$I = \frac{0.5^2}{1 * 1} = 0.25$$

To order to give equal weight to absorption and backscatter features in the band selection process, a modified spectrum for I*(λ) is defined as:

$$I^* = \begin{cases} I & for\ I > 1 \\ -\dfrac{1}{I} & for\ 0 < I < 1 \end{cases} \tag{12}$$

Using this definition, both concavities will be on the same scale and a concave-down feature (hence: backscattering) will give a positive value (>1) while a concave-up feature (hence: absorption) will give a negative value (<-1) and be treated the same numerically. For example, in the above example, the result for the new definition of the inflection or 2nd derivative estimator would be 4 and -4.

A second issue is to determine what values to assign to the upward and backward operators in the dilation filter. One could pick the optimal value for the inflection using all possible combinations of m and n. The problem with this method is that when m and n are large, the difference between the channels for which the inflection is calculated and the one to which it is compared can be influenced by the signal to noise ratio being at the low and high wavelengths in a typical camera/spectrograph system. Thus the resulting optimal regions selected can be scientifically or physically difficult to explain. Thus a limit is placed on the maximum value of the m, n operators from a practical point of view. The minimal value of m, n is 1. Thus, one can select the optimal range of the m and n wavelet filter operators (either a maximum (backscattering) or a minimum (absorption) for all combinations of m and n between 1 and the maximal value (in this paper this maximal value used was selected as 7). The resulting derivative estimator spectra (inflection spectra) using equation 12 was calculated and is shown below, using the previously shown BRDF spectra shown in Figure 8 above.

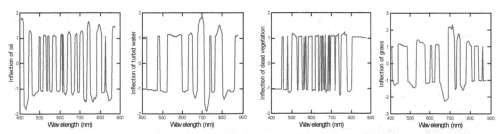

Fig. 11. BRDF Inflection spectra using the reflectance spectrums above. From left to right: an oil film (1mm thick) on clear water, turbid water, dead vegetation (dead leaves) and grass.

The inflection algorithm can also be applied to the contrast spectrums (to enhance variation in the contrast spectrum). The result of this calculation is given in the following figures.

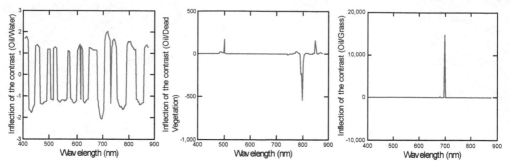

Fig. 12. Inflection of the contrast spectra. The contrast target is weathered oil with different backgrounds. From left to right: turbid water, dead vegetation (dead leaves) and grass are the contrast backgrounds.

Once the inflection spectra are calculated, it is also possible to apply Weber's definition of the contrast to the inflection spectra instead of the BRDF. The resulting contrast spectrums are given in the following figure:

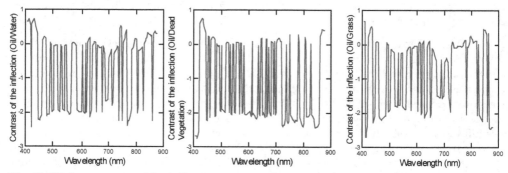

Fig. 13. Weber contrast of the inflection spectra. The target is weathered oil with different backgrounds. From left to right: turbid water, dead vegetation (dead leaves) and grass. Optimal bands and spectral regions are indicated by the greatest positive or negative values across the spectrums.

The result of the optimization procedures yields a band or band ratio for the different types of contrast (Weber's contrast, contrast of the inflection or inflection of the contrast). The optimal bands using the different techniques that were obtained using weathered oil film as the target and water, dead vegetation or grass as backgrounds are shown in the Table 1, and are used in processing hyperspectral imagery collected using the methods in the results section of this paper (see Table 1).

7. Collection of hyperspectral imagery from littoral zone

In order to detect and discriminate the presence of weathered oil on a near shore habitat or the spatial extent of weathered oiled along a shoreline, a novel and new technique has been developed for collecting HSI imagery from a small vessel (anchored or underway), or the sensor mounted in the littoral zone. The resulting HSI imagery produces pixel sizes or ground sampling distances (GSD) on the order of several mm to cm scales, depending upon the distance between the sensor and the shoreline. The purpose of collecting this type of imagery is to (1) reduce atmospheric affects and (2) minimize the influence of the "mixed pixel" and "adjacency effects" in selecting spectral regions for detection of weathered oil and for testing algorithms. The results are also immediately and directly applicable to low altitude airborne imagery, especially if the same sensor is used aboard the airborne platform.

	Water	Mud and oil	Sand	Vegetation
Weber's contrast	$Band(\lambda = 759\text{nm})$	$Band(\lambda = 378\text{nm})$	$Band(\lambda = 368\text{nm})$	$Band(\lambda = 345\text{nm})$
Inflection of the contrast	$Band(\lambda = 382\text{nm})$	$Band(\lambda = 360\text{nm})$	$Band(\lambda = 382\text{nm})$	$Band(\lambda = 710\text{nm})$
Contrast of the inflection	$Band(\lambda = 420\text{nm})$	$Band(\lambda = 684\text{nm})$	$Band(\lambda = 424\text{nm})$	$Band(\lambda = 684\text{nm})$
Multiple wavelength contrast	$\dfrac{Band(\lambda = 454nm)}{Band(\lambda = 345nm)} - 1$	$\dfrac{Band(\lambda = 751nm)}{Band(\lambda = 345nm)} - 1$	$\dfrac{Band(\lambda = 751nm)}{Band(\lambda = 345nm)} - 1$	$\dfrac{Band(\lambda = 751nm)}{Band(\lambda = 345nm)} - 1$
Multiple wavelength contrast of the inflection	$Band(\lambda = 394nm)$ $-Band(\lambda = 363nm)$	$Band(\lambda = 394nm)$ $-Band(\lambda = 363nm)$	$Band(\lambda = 394nm)$ $-Band(\lambda = 363nm)$	$Band(\lambda = 394nm)$ $-Band(\lambda = 363nm)$

Table 1. Resulting bands or band ratios for the optimization of: the contrast (Weber's definition), the inflection of the contrast, the contrast of the inflection spectra, the multiple wavelength contrast (as defined above) and the multiple wavelength contrast of the inflection spectra. In each case, weathered oil is the target and the background is: water, mixture of oil and mud, sand or vegetation.

The sensor used to view the shoreline can be directly mounted on the vessel or can be mounted above the water but near the shore using a tripod or in a vessel. In the case of a mounted sensor on a vessel, the vessel is anchored at two points, allowing movement in mainly one direction (for example the boat is anchored to mainly allow motion due to waves in the pitching direction. Fixed platform mounting does not require motion correction, however the data collected from the anchored vessel requires roll motion correction (in this case pitch correction).

In order to perform this correction, an IMU (inertial measurement unit) is attached to the HSI sensor and collects the sensor motion information while the pushbroom sensor sweeps or is rotated (using a rotation stage) along the shoreline being investigated. This correction will be applied before any further processing of the contrast algorithms are applied to the imagery taken in the Northern Gulf of Mexico and shown below. An example of the measurement scheme that has been used to detect and discriminate weathered oil (as described above) is shown below.

The image below (right) is the resulting hyperspectral image 3 band RGB display of a shoreline that has been impacted by a recent oil-spill in the Gulf of Mexico region, near Bay Jimmy, Louisiana.

Fig. 14. The HSI imaging system (left) is placed upon a small vessel or a fixed platform (tripod) in shallow water types within viewing distance of a shoreline. The sensor sweeps the shoreline and the pushbroom sensor produces a hyperspectral image of the shoreline as shown in the above HSI 3 band image (right). Note the ability to see gravity and capillary waves, small grasses on the shoreline as well as weathered oil at the land-water margin. Image collected February 28, 2011 in Barataria Bay, Louisiana

In this case a vessel mounted sensor was used and the image was corrected for the platform motion (right). To illustrate the influence of the motion of a small vessel, and the necessary IMU corrections needed, a shoreline was imaged from a vessel (below left image) and from a fixed *in-situ* platform (right image) in April 2011 the platform motion (right).

Fig. 15. A hyperspectral image (left) 3 band RGB display of a littoral zone using a pushbroom sensor mounted on a vessel anchored at two points. During the acquisition of the hyperspectral image the sensor records the pitching effect of the anchored vessel that needs to be corrected using an IMU sensor due to the water surface gravity waves. The influence of this motion can clearly be seen in the image if no correction is applied (left). The shoreline area (right) acquired when the pushboom sensor was mounted on fixed platform above the water. In this case no correction needs to be applied to the image. Note the clarity of the water surface capillary and small gravity waves.

8. Conclusion

The purpose of this paper has been to describe different calibration approaches and techniques useful in the development and application of remote sensing imaging systems. Calibration includes the use of laboratory and field techniques including the scanning of photogrammetric negatives utilized in large format cameras, as well as *in-situ* targets and spectral wavelength and radiance calibration techniques. A newly integrated hyperspectral airborne pushbroom imaging system has been described in detail. Imagery from different integrated imaging systems were described for airborne remote sensing algorithm developments using high spatial resolution (on the order of a few mm² to larger sub meter pixel sizes) imaging systems. The high spatial and spectral resolution imagery shown in this paper are examples of technology for characterization of the water surface as well as subsurface features (such as weathered oil) in aquatic systems.

Other ongoing applications in the Marine & Environmental Optics Lab making use of data from the remote sensing systems described in this paper are (a) land surface vegetation studies needed for ongoing climate change studies currently being conducted in coastal Florida scrub vegetation studies and (b) layered radiative transfer modeling of surface and subsurface oil signatures for sensor comparisons and related algorithm development to detect surface and subsurface oil using spectral and spatial data fusion and sharpening techniques.

9. Acknowledgments

The work presented in this paper has been supported in part by the Northrop Grumman Corporation, NASA, Kennedy Space Center, KB Science, the National Science Foundation, the US-Canadian Fulbright Program, and the US Department of Education, *FIPSE* & European Union's grant *Atlantis STARS* (Sensing Technology and Robotics Systems) to Florida Institute of Technology, the Budapest University of Engineering and Economics (BME) and the Belgium Royal Military Academy, Brussels, in order to support of the involvement of undergraduate students in obtaining international dual US-EU undergraduate engineering degrees. Acknowledgement is also given to recent funding from the Florida Institute of Oceanography's BP Corporation's research grant award in support of aerial image acquisition.

10. References

Aktaruzzaman, A., [Simulation and Correction of Spectral Smile Effect and its Influence on Hyperspectral Mapping]. MS Thesis, International Institute for Geo-Information Science and Earth Observation, Enschede, Netherlands, pp. 77 (2008)

Bostater, C., "Imaging Derivative Spectroscopy for Vegetation Dysfunction Assessments", SPIE Vol. 3499, pp. 277-285 (1998)

Bostater, C., Ghir, T., Bassetti, L., Hall, C., Reyier, R., Lowers, K., Holloway-Adkins, K., Virnstein, R., "Hyperspectral Remote Sensing Protocol for Submerged Aquatic Vegetation in Shallow Water", SPIE Vol. 5233, pp. 199-215 (2003)

Bostater, C., Jones. J., Frystacky, H., Kovacs, M., Joza, O., "Image Analysis for Water & Subsurface Feature Detection In Shallow Waters", SPIE Vol. 7825, pp. 7825-17-1 to 7, (2010).

4

CSIR – NLC Mobile LIDAR for Atmospheric Remote Sensing

Sivakumar Venkataraman
Council for Scientific and Industrial Research,
National Laser Centre, Pretoria
University of Pretoria, Department of Geography
Geoinformatics and Meterology, Pretoria
University of Kwa-Zulu Natal, Department of Physics, Durban
South Africa

1. Introduction

Remote sensing is a technique for measuring, observing, or monitoring a process or object without physically touching the object under observation. The remote sensing instrumentation is not in contact with the object being observed, remote sensing allows - to measure a process without causing disturbance - to probe large volumes economically and rapidly, such as providing global measurements of aerosols, air pollution, agriculture, environmental impacts, solar and terrestrial systems, ocean surface roughness and large-scale geographic features. The modern atmosphere remote sensing technique offers to study in detail, the atmospheric physics/chemistry and meteorology. In general, observation, validation, and theoretical simulation are highly integrated components of atmospheric remote sensing. Active and passive remote-sensing techniques and theories/formulation methods for measuring atmospheric and environmental parameters have advanced rapidly in recent years. Active remote sensing instrumentation includes an energy source on which the measurement is based. In this case, the observer can control the energy source and the examples of this class are RADAR, LIDAR, SODAR, SONAR etc. Passive remote sensors do not include the energy source on which the measurement is based. They rely on an external light, which is beyond the control of the observer and examples of this class are optical and radio telescopes, radiometers, photometers, spectrometers etc.

2. LIDAR as a remote sensing probe

LIDAR (LIght Detection And Ranging) is also called as "Optical RADAR" or "Laser RADAR". It is a powerful and versatile remote sensing technique for high resolution atmospheric studies. It complements the conventional RADAR for atmospheric studies by being able to probe the region not accessible to the RADAR and study micro-scales of the atmosphere. The LIDAR probing of the atmosphere started in early 1960s and pursued intensively over the past five decades. *Fiocco and Smullins* (1963) used Ruby Laser with a feeble energy of 0.5J, obtained Rayleigh scattering signals from the atmosphere upto 50 km altitude and also detected dust layers in the atmosphere. *Ligda* in 1963 made the LIDAR measurements of cloud heights in the troposphere. Recent developments leading to the availability of more powerful, relatively rugged and highly efficient solid state lasers and improvements in detector technology as well as data acquisition techniques have resulted, LIDARs as a potential tool for atmospheric studies. Both continuous wave and pulsed laser systems have been extensively used and they are currently operational for the study of atmospheric structure and dynamics, trace constituents, aerosols, clouds as well as boundary layer and other meteorological phenomena. Currently laser systems are being used for probing the atmosphere begin from surface (near boundary layer) to lower thermosphere altitudes (upto ~100 km).

2.1 LIDAR principle

LIDAR is one of the most powerful remote sensing techniques to probe the earth's middle atmosphere. The basic principle of probing the atmosphere by LIDAR is similar to that of the RADAR. In the simplest form, LIDAR employs a laser as a source of pulsed energy of useful magnitude and suitably short duration. Typically Q-switched ruby (wavelength=0.69 μm) or Neodymium (wavelength 1.06 μm) laser systems are used to generate pulses having peak powers measured in tens of megawatts in the duration of 10-20 nsec. Pulses with such energy (i.e. of the order 1 joule) are directed in beams by suitable optical systems. The advantage of laser, as it has specific properties of virtually monochromatic and highly coherent and collimated.

As the transmitted laser energy passes through the atmosphere, the gas molecules and particles or droplets cause scattering. A small fraction of this energy is backscattered in the direction of the LIDAR system and is available for detection. The scattering of energy in directions other than the direction of propagation, or absorption by the gases and particles, reduces the intensity of the beam, which is said to be attenuated. Such attenuation applies to both the paths (to and fro) of the distant backscattering region.

The LIDAR backscattered energy is collected in a suitable receiver by means of reflective optics and transferred to a photo-detector (commonly referred to a photo-multiplier). This produces an electrical signal, the intensity of which at any instant is proportional to the received LIDAR signal power. Since the light travels at a known velocity, the range of the scattering region produces the signal received at any instant can be uniquely determined from the time interval of the sampled signal from the transmitted pulse. The magnitude of the received signal is determined by the backscattering properties of the atmosphere at successive ranges and by the two-way atmospheric attenuation. Atmospheric backscattering intern depends upon the wavelength of the laser energy used, and the number, size, shape and refractive properties of the particles (droplets and molecules) intercepting the incident energy. Backscattering from an assemblage of scatterers is a complicated phenomenon; in general, the backscattering increases with increasing scatterer concentrations.

The electrical signal from the photo detector thus contains information on the presence, range and concentration of atmospheric scatterers. Various forms of presenting and analyzing such signals are available. In the simplest form they may be presented on an oscilloscope in a coordinate system showing received signal intensity as a function of range. Since such signals are transient, (1 km of range is represented by an interval of time of ~7 μs), it is necessary to photograph several such oscilloscope displays to obtain adequate data for presentation.

Figure 1 shows the schematic diagram of LIDAR probing of the atmosphere in which P_0 represents the laser-transmitted pulse energy. Let us consider at an altitude z the scattering take place, hence a factor T attenuates the intensity of light pulse. The radiation scattered in backward is $P_0 T \beta$, where β is the backscattering coefficient (sum of Rayleigh scattering by air molecules and Mie scattering by aerosol particles). Since the backscattered radiation travels the same distance r before being detected by the telescope, it further undergoes attenuation by the same factor T. Thus the intensity of the backscattered signal detected at the telescope becomes $\dfrac{P_0 T^2 \beta A}{r^2}$, where A is the area of the telescope receiving the backscattered radiation.

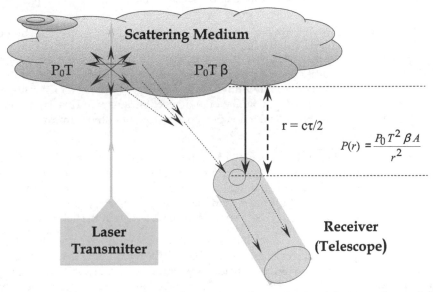

Fig. 1. Schematic diagram showing the basic principle involved in LIDAR probing of the atmosphere.

LIDARs may be configured into two ways; (a) Mono-static configuration in which both transmitter and receiver are collocated. (b) Bi-static configuration in which both transmitter and receiver are separated by some distance.

2.2 LIDAR equation

The transmitted laser beam gets scattered in all directions at all altitudes, the backscattered echoes are received by the telescope and their intensities are measured. The field of view of the telescope is kept larger than beam divergence, in order to accommodate the beam completely at all altitudes. The received signal intensity is described in terms of the LIDAR equation as given by (Fiocco, 1984);

$$P(r) = P_0 \, \eta \left(\frac{A}{r^2}\right) \left(\frac{c\tau}{2}\right) \beta(r) \exp\left[-2\int \alpha(r)dr\right] \qquad (1)$$

Where P(r) is the instantaneous power received at time t from an altitude (range) r, P_0 is the transmitted power, η is the system constant which depends on the transmitter and receiver efficiencies. A is the area of primary (collecting) mirror of the receiving telescope. The term $\left(\dfrac{A}{r^2}\right)$ is the solid angle subtended by the primary mirror at the range r. This simple expression for solid angle is applicable for monostatic only because all the transmitted energy contributes to the backscattered signal from the range r. The term $\left(\dfrac{c\tau}{2}\right)$ gives the length of the illuminated path, which contributes to the received power, where c is the velocity of light and τ is the pulse duration of the laser beam.

The $\left(\dfrac{c\tau}{2}\right)$ term determines the minimum spatial resolution available in the direction of the beam propagation. In the transverse direction the spatial resolution depends on the laser beam width at particular altitude. In a typical LIDAR system the pulse duration of the laser beam is of the order of few nanoseconds and the beam divergence is less than a milli-radian, which corresponds to a scattering volume of a few cubic meters. This is the greatest advantage of the LIDAR technique which is not possible by any other atmospheric remote sensing technique.

$\beta(r)$ is the volume backscattering coefficient of the atmosphere at range r. It gives the fractional amount of the incident energy scattered per steradian in the backward direction per unit atmospheric path length and has the dimension of $m^{-1}sr^{-1}$. α is the volume attenuation coefficient of the atmosphere and has the unit of m^{-1}, defined as twice the integral between the transmitter and the scattering volume to obtain the net transmission.

The term α and β include the contribution from air molecules, aerosols and the other atmospheric species. The problem related with the LIDAR equation is that it contains two unknowns, α and β, which make it difficult to obtain the general solution. Appropriate inversion methods (*Fernald et al.*, 1984; *Klett*, 1981 & 1985) have been developed to solve the equation. The LIDAR equation however assumes only single scattering. Contribution arising from multiple scattering is important for high turbidity cases such as clouds and fogs.

2.3 LIDAR scattering / absorption mechanisms

As the radiant energy passes through the atmosphere it undergoes transformations like absorption and scattering. Absorption (or emission) of radiation takes place when the atoms or molecules undergo transition from a energy state to another. Scattering is the deflection of incoming solar radiation in all directions. Scattering of radiation depends to a large extent on particle size. There are several scattering / absorption mechanisms that occur when the laser energy interacts with the atmosphere. The predominant scattering is quasi-elastic scattering from aerosols (Mie scattering) or molecules (Rayleigh Scattering). The quasi-elastic nature arises from the motion of the molecules or aerosols along the direction of the laser beam. Aerosols, generally move with the air mass, give rise to smaller Doppler shifts, while the molecules, move at high speed, give rise to larger Doppler shifts. Another form of atmospheric elastic scattering is resonance fluorescence. In-elastic scattering includes Raman Scattering and Non-Resonance Fluorescence. These scattering processes, sometimes in combination with molecular absorption, form the basis for various types of LIDAR remote sensing techniques. The most well known is DIAL (DIfferential Absorption LIDAR) or DASE (Differential Absorption Scattering Energy). Table 1 summarizes these mechanisms.

Technique	Atmospheric measurements
Rayleigh Scattering	Air Density and temperature (above 35 km)
Mie Scattering	Cloud, Smog, Dust, Aerosols (Below 35 km)
Raman Scattering	N_2, CO_2, H_2O and Lower Atmosphere temperature (less than 20 km)
Differential Absorption LIDAR (DIAL)	Trace Species, like O_3, NO_2, CO_2, CH_4, CO, H_2O (for upto 50 km)

Table 1. Main scattering / absorption process of laser-atmosphere Interactions.

2.3.1 LIDAR scattering / absorption mechanisms

Rayleigh scattering

In 1890's Lord Rayleigh showed that the scattering of light by air molecules is responsible for the blue color of the sky. He showed that, when the size of the scatterer is small compared to the wavelength of the incident radiation. Rayleigh scattering mainly consists of scattering from the atmospheric gases. This type of scattering is varies nearly as the inverse of fourth power of interactive wavelength and directly proportional to sixth power of the radius of the scatter.

Mie scattering

When the sizes of the scattering particles are comparable to or larger than the LIDAR wavelength, the scattering is governed by Mie theory. Pollen, dust, smoke, water droplets, and other particles in the lower portion of the atmosphere cause Mie scattering. Mie scattering is responsible for the white appearance of the clouds. Note that for a given incident wavelength as the size of the scatterer is reduced, the scattering computed using Mie theory coincides with the results obtained using Rayleigh formula. Thus Rayleigh scattering is said to be a special case of Mie scattering. The Mie scattering is directly proportional to wavelength and proportional to the volume of the scatterers.

Raman scattering

Raman scattering is the process involving an exchange of a significant amount of energy between the scattered photon and the scattering species. Thus the Raman scattering component is shifted from the incident wave frequency by an amount corresponding to the internal energy of the species. The Raman scatter has both down-shifted (stokes) and up-shifted (anti-stokes) lines in its spectrum. The cross section for Raman scattering is small and compared to Rayleigh scattering, it is smaller by about three orders of magnitude. However, by LIDAR technique, it offers a valuable means for identifying and monitoring atmospheric constituents and also for temperature measurements in the lower atmosphere. The technique makes use of stokes line since its intensity is much greater than that of anti-stokes line.

Differential absorption technique

The most sensitive and effective absorption method for the measurement and monitoring of air pollutants is the "Differential Absorption LIDAR (DIAL)" technique. In this technique, the pulsed laser transmitter emits signals at two wavelengths, λ_{on} and λ_{off} corresponds to absorption line and other outside the absorption line. The received backscatter power on and off wavelength is given by

$$P_{on} = \frac{E_{on}\beta_{on}(r)C}{2r^2}\exp\left[-\int_0^r 2\alpha_{on}(r')dr'\right] \tag{2}$$

$$P_{off} = \frac{E_{off}\beta_{off}(r)C}{2r^2}\exp\left[-\int_0^r 2\alpha_{off}(r')dr'\right] \tag{3}$$

Where P is the received backscatter power at time $t = 2r/c$, r is range, E is the transmitted laser pulse energy, β is the atmospheric backscatter coefficient, α is the atmospheric extinction coefficient and C is system constant. The atmospheric absorption and extinction coefficient can be expressed interms of aerosol and molecular components.

In this method, the ratio of the received backscattered power between λ_{on} and λ_{off} wavelength is directly proportional to the number concentrations of the molecule/gaseous pollutants.

Table-2 provides the primary laser sources, which are used for atmospheric applications. In which solid-state lasers are popular. The first laser systems used with the flash lamp pumped is a Q-switched ruby laser. Now it has been implemented in Nd-YAG laser system also.

Laser	Wavelength	Energy per pulse	Efficiency (%)
Ruby	0.694 µm	2-3 J at 0.5 Hz	0.1 – 0.2
Nd:YAG	1.06 µm	1 J at 10 Hz, 10 ns pulse	1 - 2 *
CO_2	9-11 µm multi-line	1-10 J at 1-50 Hz	10 – 30
CO_2	Tunable	0.1 J at 10 Hz	5
CO	5 – 6.5 µm	Not very popular for pulsed operation	10
Dye lasers Flash lamp pumped	0.35-1.0 µm	0.1 – 20 J	1

*Note: More recently using diode array pumping more than 20 % efficiencies have been achieved.

Table 2. Primary laser sources used for atmospheric applications

2.4 Applications of LIDAR

LIDARs are used in variety of applications in the field of atmospheric science. Some of the main applications are outlined, below.

LIDAR for the aerosol studies

The LIDAR provides measurements of the optical backscattering cross section of air as a function of range and wavelength. This information may be subsequently interpreted to obtain profiles of the aerosol concentration, size distribution, refractive index, scattering, absorption and extinction cross sections and shape. The scattering involves with aerosol is mainly due to Mie scattering. Details on Mie scattering are provided in the earlier section.

LIDAR for the cloud studies

LIDARs are well suited and widely used for determining the characteristics of clouds, especially high altitude clouds, because of their high range resolution and high sensitivity to hydrometeors. The sharp enhancement in the Mie backscattered LIDAR signal makes possible the detection and characterization of the clouds (*Fernald*, 1984). Although one channel LIDAR can define physical boundaries of clouds, polarization diversity gives fundamental principles to distinguish between water and ice phase of the clouds. The LIDAR measurements of scattering ratio and linear depolarization ratio (LDR) provide the cloud parameters and information on the thermodynamic phase of the cloud particles.

LIDAR to determine middle atmospheric temperature

In the height range, where the contribution from the Mie backscatter is negligible ($\geq$ 30 km), the recorded signal is due to the Rayleigh backscatter and its intensity, corrected for the range and atmosphere transmission, is proportional to the molecular number density. Using the number density taken from an appropriate model for a specified height, where the signal-to-noise ratio is fairly high, the constant of proportionality is evaluated and thereby the density profile is derived. Taking the pressure at the top of the height range (say 90 km) from the atmospheric model, the pressure profile is computed using the measured density profile, assuming the atmosphere to be in hydrostatic equilibrium. Adopting the perfect gas law, the temperature profile is computed using the derived density and pressure profiles. The analysis closely follows the method described by *Hauchecorne and Chanin* (1980).

LIDAR to determine the wind speed

Doppler LIDARs make use of the small change in the operating frequency of the LIDAR due to motion of the scatterers to measure their velocity. Using the technique called heterodyning, the returned backscattered signal is used with another laser beam so that they interfere, yielding a more easily measurable signal at radio wave frequency. The frequency of the radio wave will be equal to the difference between the frequencies of the transmitted and the received signals. The application of Doppler LIDAR in atmospheric remote sensing is to measure wind velocity, i.e., wind speed and direction in addition to other parameters.

LIDAR for the measurements of vertical profile of ozone

The DIAL technique has been used to provide vertical profiles of the ozone number density from ground to 40-50 km height level. The basic principle of the DIAL technique is described in earlier section (section 2.3.1). In this technique, the laser transmitter emits signals at two close wavelengths, λ_{on} and λ_{off} corresponding to a peak and trough, respectively in the absorption spectrum of the species of interest. The ratio of the two received signals due to backscattering corresponds to the absorption produced by the species (O_3) in the range cell defined by the laser pulse duration and receiver gate. The amount of absorption is directly related to the concentration of the constituent.

LIDAR for lower atmospheric temperature and minor constituents

Raman LIDAR is useful in obtaining molecular nitrogen concentration from low altitudes (below 30 km) where Rayleigh LIDAR technique is not applicable due to the presence of aerosols. In case of Raman scattered signal the radiation emerging only from the N_2 molecules are detected that is proportional to the number density of air molecules. Temperature could be derived from the number density as the case of Rayleigh LIDAR. Raman scattering is also used to detect different molecular species present in the atmosphere.

LIDAR in space

Ground-based LIDAR provides atmospheric data over a single viewing site, while LIDAR aboard an aircraft can gather data over an area confined to a region. Thus the ground-based and airborne LIDARs provide data over a limited area of a specified region of the earth. Space borne (satellite-based) LIDARs, on the other hand, have the potential for collecting data on a global scale, including remote areas like the open ocean, in a short period of time.

3. Lidar activities in south africa

Although ground-based LIDAR systems exist in many developed countries and largely concentrated in northern hemisphere mid- and high latitude, it is still a very novel technique for South Africa and African countries. A recent survey on the available LIDAR system around the world, noticed that there are currently two different LIDARs available in South Africa, located in Pretoria and Durban (see. Figure 2). Both LIDAR systems are similar in operation and different in specifications and the objectives of measurements. The Durban LIDAR is operated at University of KwaZulu-Natal as part of cooperation between the

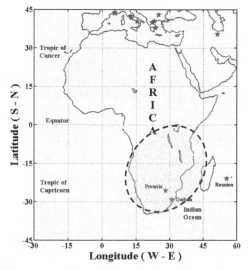

Fig. 2. Geo-graphical position of LIDAR sites in Pretoria and Durban.

Reunion University and the Service d'Aéronomie (CNRS, IPSL, Paris) for atmosphere research studies, especially to study the upper troposphere and lower stratosphere (UTLS) aerosol structure and middle atmosphere temperature structure and Dynamics. The Council for Scientific and Industrial Research (CSIR) National Laser Centre (NLC) in South Africa has recently designed and developed a mobile LIDAR system to contribute to lower atmospheric research in South Africa and African countries. The CSIR mobile LIDAR acts as an ideal tool to address atmospheric remote sensing measurements from ground to 40 km and to study the atmosphere aerosol/cloud studies over Southern Hemisphere regions and this will encourage collaboration with other partner's in-terms of space-borne and ground based LIDAR measurements.

4. CSIR - NLC mobile LIDAR system

4.1 System description

The CSIR NLC mobile LIDAR has been configured into mono-static that maximizes the overlap of the outgoing beam with the receiver field of view. The LIDAR system has been mounted in a mobile platform (van) with a special shock absorber frame. Figure 3 shows a 3-D pictorial representation of the mobile LIDAR with 2-D scanner. In general, any LIDAR systems can be sub-divided into three main sections, a laser transmitter, an optical receiver and a data acquisition system.

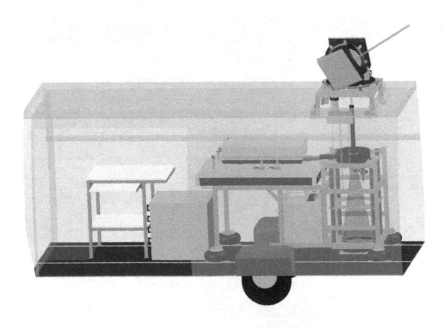

Fig. 3. A 3-D pictorial representation of the CSIR-NLC mobile LIDAR with 2-D scanner

The important main specifications of the LIDAR system are listed in Table 3.

Parameters	Specifications
Transmitter	
Laser Source	Nd:YAG - Continuum®
Operating Wavelength	532 nm and 355 nm
Average pulse energy	120 mJ (at 532 nm) 80 mJ (at 355 nm)
Beam Expander	5 x
Pulse width	7 ns
Pulse repetition rate	10 Hz
Beam Divergence	0.12 mrad after Beam Expander
Receiver	
Telescope type	Newtonian
Diameter	404 mm
Field of View	0.5 mrad
PMT	Hamamatsu® R7400-U20
Optical fibre	Multimode, 600 μm core
Filter FWHM	0.7 nm
Signal and Data Processing	
Model	Licel® TR15-40
Memory Depth	4096
Maximum Range	40.96 km
Spatial Resolution	10 m
PC	
TR15-40 Interface	Ethernet
Processor	Intel® Core2Duo 2.6 GHz
Operating system	Windows® XP Pro
Software Interface	NI LabVIEW®
Application	
Aerosol/Cloud study	0.5 km to 40 km
Water Vapour	0.5 km to 12 km (to be done)
Temperature	0.5 km to 20 km (to be done)
Scanner resolution (minimum)	
X-axis (Horizontal)	0.002 rad
Y-axis (Vertical)	0.001 rad

Table 3. Major specifications of the CSIR-NLC mobile LIDAR system

4.1.1 Laser transmission

The transmitter employs a Q-Switched, flash lamp pumped Nd:YAG (Neodymium (Nd) impurity ion concentration in the Yttrium Aluminum Garnet (YAG)) solid-state pulsed laser (Continuum®, PL8010). Nd:YAG lasers operate at a fundamental wavelength of 1064 nm. Second and third harmonic conversions are sometimes required, depending on the application and are accomplished by means of suitable non-linear crystals such as Potassium (K) Di-hydrogen Phosphate (KDP). At present, the second (532 nm) and third (355 nm) harmonic is utilized and the corresponding laser beam diameter is approximately 8 mm. The laser beam is passed through a beam expander (expansion of 5 times), before being sent into the atmosphere, thereby the beam divergence is reduced by the factor of 5 (i.e, 0.6 mrad to 0.12 mrad). The resultant expanded beam has a diameter of 40 mm and is then reflected upward using a flat, 45 degree turning mirror. The entire transmission setup is mounted on an optical breadboard. The power supply unit controls and monitors the operation of the laser. It allows the user to setup the laser's flash lamp voltage, Q-Switch delay and the laser repetition rate. It also monitors system diagnostics such as the flow and temperature interlocks. The power supply also incorporates a water to water heat exchanger which regulates the temperature and quality of water used to cool the flash lamps and laser rods. The inbuilt laser Control Unit (CU601) provides cooling group interlocks, which sense water temperature, water level and water flow. A cooling group interlock violation halts the laser operation and reports the interlock violation to the remote box. At present, the laser is being utilized at the pulse repetition rate of 10 Hz.

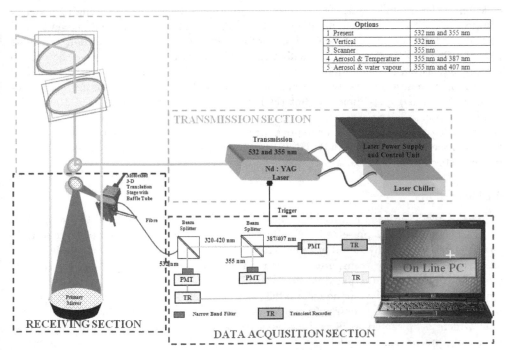

Options		
1	Present	532 nm and 355 nm
2	Vertical	532 nm
3	Scanner	355 nm
4	Aerosol & Temperature	355 nm and 387 nm
5	Aerosol & water vapour	355 nm and 407 nm

Fig. 4. Block diagram of CSIR-NLC mobile LIDAR illustrating different components.

4.1.2 Receiver section

The receiver system employs a Newtonian telescope configuration with a 404 mm primary mirror. The backscattered signal is first collected and focussed by the primary mirror of the telescope. The primary reflecting mirror has a 2.4 m radius of curvature and is coated with an enhanced aluminium substrate. The signal is then focused toward to a secondary 45 degree plane mirror and coupled into an optical fibre. One end of the fibre is connected to an optical baffle which receives the return signal from the telescope. The other end is connected to an optical tube with collimation optics and the PMT. We have also employed a motorized 3-Dimensional translation stage in order to accurately align the fibre using PC control.

4.1.3 Data acquisition

PMT is used to convert the optical backscatter signal to an electronic signal. The PMT is installed in an optical tube and is preceded by a collimation lens and narrow band pass filter. The PMT used is a Hamamatsu R7400-U20. It is a subminiature PMT which operates in the UV to NIR wavelength range (300 nm – 900 nm) and has a fast rise time response of 0.78 ns. It is specially selected for minimal noise with an anode dark current.

Data acquisition is performed by a Licel transient recorder (TR). The system is favored by its dual capability of simultaneous acquiring analog and photon count signal, which makes it highly suited to LIDAR applications by providing a higher dynamic range. The TR15-40 is the model that was procured. It is capable of 15 MHz sampling and has a memory depth of 4096 bins. The photon count channel uses a high pass filter to select the high frequency component (>10 MHz) of the amplified PMT signal. The filtered component is then passed through a fast discriminator (250 MHz) and counter enabling the detection of single photons. The Licel system together with a LabVIEW software interface allows the user to acquire signals without any immediate programming. As mentioned earlier, the Licel data acquisition system incorporates electronics which is capable of simultaneous acquisitions of Analog Data (AD) and Photon Count (PC) data with a range resolution of 10 m. The combination of PC and AD electronics greatly extends the dynamic range of the detection channel allowing the reduction or removal of neutral density filters, which in turn greatly improves the Signal-to-Noise Ratio (SNR). The measurements are usually done at night to minimize back-ground noise.

4.2 Illustration

In general, the laser beam is directed vertically upward into the sky as depicted in figure-3. The corresponding day presented a cloudy sky and there was a passage of high-altitude cirrus, which is normally found at upper altitudes from 6 km to 15 km. Since these clouds are generally optically transparent, depend upon the physical property, laser light is passed/prevented from passing through. The observations were carried out for approximately four and a half hours and the presence of clouds is clearly seen in the height-time-backscattered signal returns for both the Analog Data (AD) and Photon Count (PC) data which is presented in Figs. 5 and 6 respectively. The figures were obtained after modifying the provided Licel–LABVIEW software, in-house, to display an automatically updated height-time-backscatter colour map in real time. The advantage of such a program

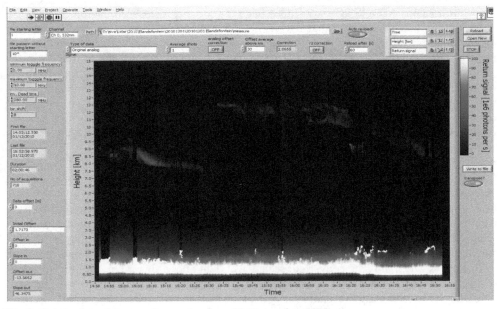

Fig. 5. Original analog signal measured on 01 December 2010

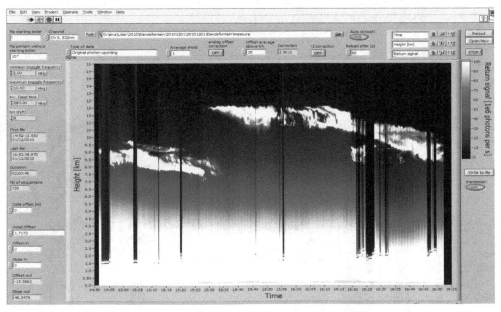

Fig. 6. Same as fig. 5 but represents the original photon count signal

is that it allows the user to infer the data simultaneously while the LIDAR system is in operational. The display can be easily visualized and the available settings enable either the AD or the PC data to be displayed, as required.

The simultaneous AD and PC acquisitions have been post processed to merge or 'glue' the datasets into a single return signal. The combined AD and PC signals allow us to use the analog data in the high signal to noise ratio (SNR) regions and the PC data in the low SNR regions. Since the output from the AD converter is voltage (V) and the output from the photon counter is counts or count rates (MHz) a conversion factor between those outputs needs to be determined in order to convert the analog data to "virtual" count rate units. First the PC data is corrected for pulse pileup using a non-paralyzable assumption (dead-time correction). The dead time corrected PC data is then determined based on the linear relationship with Analog Signal, i.e., PC = a * AD + b, over a range where the PC data responds linearly to the AD and where the AD is significantly above the inherent noise floor. The linear regression has been applied to determine the gain and offset coefficients (gluing coefficients), a and b. Thereafter, the coefficients are used to convert the entire AD profile to a "virtual/scaled" photon count rate. This is referred to as the scaled analog signal. i.e., the term, "a * AD" (see. Figure 7) and the term 'b' stands for the bin shift (offset). Commonly, the typical range is determined from the data above the threshold signal and where the PC data (see. Figure 8) is between 0.5 MHz and 10 MHz. The combined or glued signal then uses the dead-time corrected PC data for count rates below some threshold (typically 10 MHz) and the converted/scaled AD data above this point. Figure 9 displays the glued data for the above presented case (see Figure 7 and 8). Here, the gluing is performed after obtaining the dead time corrected photon count (dead time is 3.6 n sec) and also adjusting a minute bin shift between the AD and PC. The bin shift is basically a delay measured in bins (corresponding to 10 m per bin) which occurs due the detection electronics. Filters in the pre-amplifier electronics results in a delay of the AD signal with respect to the PC signal. The analog to digital conversion process also may also cause any further delay.

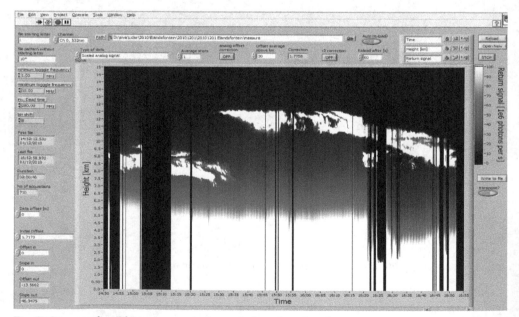

Fig. 7. Same as fig. 5 but represents the scaled analog signal

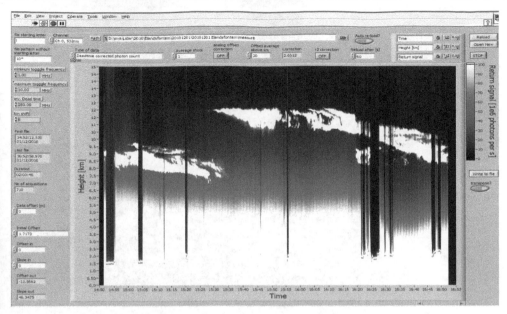

Fig. 8. Same as fig. 5 but represents the deadtime corrected photon count signal

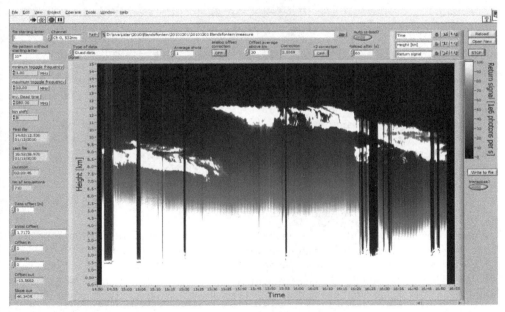

Fig. 9. Same as fig. 5 but represents the glued photon count signal

To address the dynamic range of the instrument, the range corrected glued signal (i.e., signal multiplied by R^2) is presented in figure 10. i.e., the figures represented here are the raw data multiplied by the square of the altitude, commonly referred as range corrected

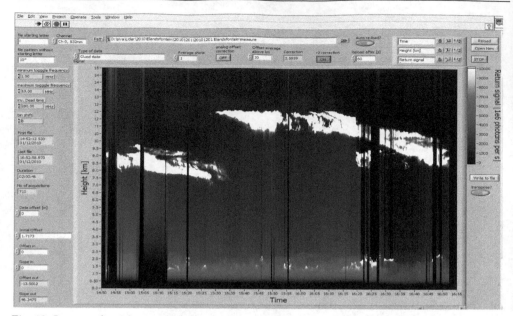

Fig. 10. Same as fig. 9 but represents the Range Corrected glued photon count signal

signal. The range corrected signal provides an equilibrium condition to the LIDAR transmitted and received backscatter signal (see. Equation 1).

The figures clearly distinguish the cloud observation from normal scattering from background particulate matter. Sharp enhancements are observed around 7.5 km and above (~12 km) indicating the presence of cloud. Such type of cloud otherwise termed as CIRRUS. The advantage of using LIDAR, is to observe the cloud thickness in addition to the cloud height. This is one of the important advantages of LIDAR measurements, in comparison with any other remote sensing measurement techniques. The advantage of having high resolution data (10 m) further address the accurate detection of cloud height and thickness, which is important for studying the cloud morphology. Apart from it, the above measurements illustrate the dynamic range of the LIDAR signal upto 35 km (though the figure is presented here upto 15 km). During the day-time measurements, to avoid the background light signal, neutral density (ND) filters are employed which protect further the PMT saturation and to investigate the maximum return signal strength.

The parameter, SNR judge always any instrument capability. Here, we have determined for the mobile LIDAR based on transmitting and receiving signal with and without emitting the LASER beam. The results are obtained by operating the LIDAR on a clear sky with the laser is being ON (Signal) and OFF (Noise) for an about twelve minutes in each cases (see Figure 11a). Figure 11(a) illustrates the temporal evolutions of LIDAR signal returns when the laser is ON and OFF. While the laser was on (first twelve minutes), a large photon count signal was obtained and when the laser was switched off (next twelve minutes), random noise photons are observed due to the background scattering from the atmosphere.

The above individual observational data are then averaged temporally and presented as a height profile of photon count in Figure 11b. Figure represents both the signal (blue) and

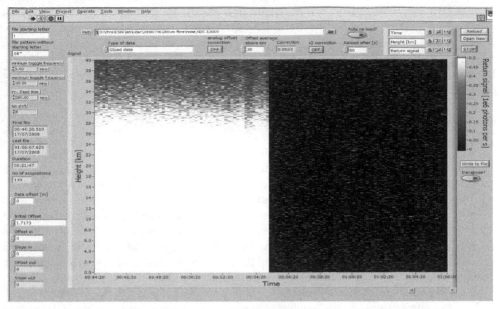

Fig. 11a. Temporal evolution of the return signal while LASER is ON and OFF mode.

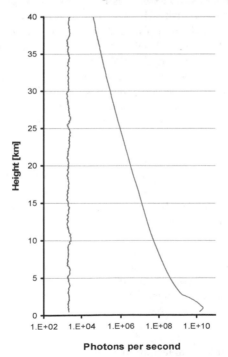

Photons per second

Fig. 11b. Height profile of averaged photon count for the above presented temporal evolution in fig. 11a.

noise (red) profiles. It is clear from the figure that the signal strength for the height region up to 40 km and shows more than 2 orders apart from the noise level. From the above results, one can conclude that the LIDAR provides reasonable measurements for the height region up to 40 km and that the signal to noise ratio is highly apart by an order of two. Further, more integration of signal may also address improvements in the SNR and the dynamic range of the instruments.

4.3 Scientific results

4.3.1 LIDAR extinction co-efficient

The altitude profiles of aerosol extinction (α) or backscatter coefficient (β) from a backscattered LIDAR signal require the solution from the LIDAR equation (see. Equation 1). As described in the LIDAR equation, the $\beta(z) = [\beta_a(z) + \beta_m(z)]$, and $\alpha(z) = [\alpha_a(z) + \alpha_m(z)]$, where, α_a and β_a are the volume extinction and backscatter coefficients of the aerosols and α_m and β_m are the volume extinction and backscatter coefficients of the air molecules. The values of α_m and β_m are calculated from the meteorological data or from a standard atmosphere model. Determinations of α_a and β_a require an inversion of the LIDAR equation. The inversion is not a straightforward process since it involves two unknowns. In this regard, a definitive relationship between the above two unknowns should be assumed. The molecular contributions to backscattering and extinction have been estimated using a reference model atmosphere (MSISE-90). This is accomplished by the normalization of the photon count with molecular density at a specified height (vary from a day to day) taken from a model (MSISE-90) and then applying the extinction correction to the backscattering co-efficient profile using iterative analysis of the LIDAR inversion equation. The estimation of aerosol backscatter co-efficient applies the downward progression from the reference altitude of ~40 km where the aerosol concentration is said to be negligible. The backscattering co-efficient profiles as computed above are also employed for the purpose of studying the cloud characteristics. For studying the aerosol concentrations, however, extinction profiles are computed by following the LIDAR inversion method as described by *Klett*, (1985).

The LIDAR inversion technique was applied to the backscattered LIDAR signal for a two continuous day measurements 30th and 31st August 2010, to determine the aerosol backscatter and extinction coefficient. Figure 12 shows the 10 minutes averaged height profile of the aerosol extinction coefficient retrieved from LIDAR signal returned on the 30th and 31st August 2010. Different height profiles for measurements on the same day are observed. It shows that the aerosols loading were not found to be stable over the measurement site. This is due to the change in the aerosol loading resulting from the change in humidity, temperature, etc. Furthermore, the differences between measurements on different day are observed. This might be due to the variations in day's background conditions, temperature, humidity, wind, cloud, solar radiation, etc.

4.3.2 Detection of cloud

Figure 13 shows an example of detection of cloud by LIDAR for the night of 23 February 2008. The laser was directed vertically upward into the sky and the corresponding night was a cloudy sky and there was a passage of cumulous clouds which is normally found at lower

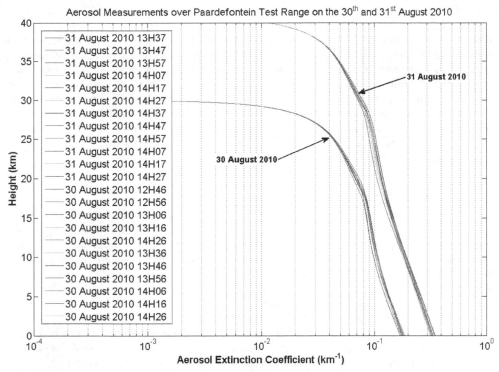

Fig. 12. Height profile of aerosol extinction coefficient retrieved from LIDAR returned signal for the 30th and 31st August 2010.

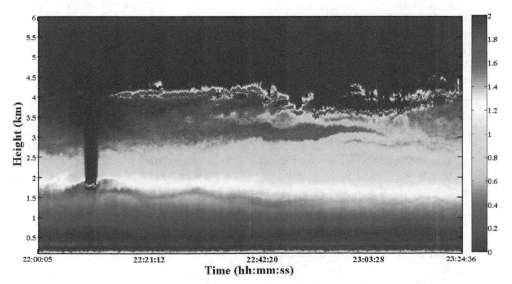

Fig. 13. Height-time-colour map of LIDAR signal returns for 23 February 2008.

height region from 3 km to 5 km. Since, these clouds are generally optically dense which prevents light to pass through. The present observations were carried out for more than two hours and the presence of clouds is clearly seen in the height-time-backscattered signal returns. Figure clearly distinguishes the cloud observation from normal scattering from background particulate matter. It shows the sharp enhancement in backscatter signal during the presence of cloud around 3.8 km and slowly has moved down to 3.5 km. The figure also demonstrates the capability of LIDAR to observe the cloud thickness (less than around 300 m) which is a unique feature of LIDAR in comparison to the satellite detection. The measured high resolution data is also important when studying cloud physics/characteristics. Otherwise, the lower height regions indicate high intensity signal returns which is due to the presence fog or aerosols.

4.3.3 Boundary layer detection

The atmospheric Boundary Layer (BL) is a part of the lower troposphere where most living beings and natural/human activities occur. It varies with space and time, and changes with height mostly during the day due to variations in solar-radiation (by several kilometers) and is quite stable over night. It is well known that the aerosol content or particulate matter in the lower atmosphere fluctuates under different background conditions (e.g., temperature, humidity and solar radiation). Such fluctuations in aerosol content, particularly the height of boundary layer, can easily be determined by means of a LIDAR (Light Detection and Ranging) backscatter signal. Based on the LIDAR backscattered signal (or/and range corrected) and by applying different criteria, one would be able to identify the boundary layer height (BLH) and thus the temporal evolution. Here, we show a typical example of deduction of BLH based on two different methods, (a) statistical and (b) slope, i.e.,

a. The statistical method applies range (z) corrected (squared) LIDAR backscattered signal (P_r), i.e., $P_r * z^2$. The BLH is identified by the height where the maximum standard deviation in the range corrected signal. Here, the mean value is obtained by the integration of consecutive 5 profiles (corresponds to 50 sec) (*Chiang and Nee*, 2006).

b. The slope method is based on the LIDAR backscattered signal (P_r) and their gradient (dP_r/dz). The identified minimum value in the slope (between P_r and dP_r/dz) defines the BLH (*Egert*, 2008).

Figure 14 shows the temporal (~2 hrs) evolution of LIDAR backscattered signal for the day of 27 May 2011. The figure is superimposed by the deducted BLH based on the two methods, statistical (Black circle) and slope (pink star). It is clear from the figure that the BLH varies significantly over time. In general, maximum BLH is found during the noon, as expected during the day that the earth's surface heats up due to solar radiation and this results in various thermodynamic chemical reactions causing turbulence in the PBL. The boundary layer height is therefore expected to vary more during the day and to stabilize after sun-set. The slope method provided a higher value in comparison with the statistical method (based on standard deviation) and the difference is found to be ~1 km. To conclude, deduction of BLH by the statistical method provides better results compared to the slope method.

4.3.4 Comparison with satellite measurements

The extinction profile derived from the LIDAR and compared /validated using ground based and satellite borne instruments. Figure 15 presents the height profile of the extinction

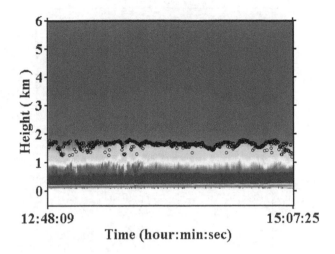

Fig. 14. Height-Time-Color map of LIDAR signal returns (arb.unit) for 27 May 2011. The figure is overlapped by the determined boundary layer height (Black: statistical method based on range corrected signal, Pink: slope method)

coefficient derived from the LIDAR data taken during the nights of 25 February 2008. The profiles are overlapped by the Stratosphere Aerosol Gas Experiment (SAGE-II) extinction data at 525 nm collected over southern Africa regions (Latitude, 15°S to 40°S and 10°E to 40°E and Longitude). The extracted mean aerosol extinction coefficients are from version 6.20 series of ~21 years (1984-2005). Here, we have used the corresponding monthly-mean extinction profiles (February). We have considered the SAGE-II profile as far as possible above 3-4 km, keeping in mind that the lower height region measurements are inaccurate due to a low signal to noise ratio (SNR) (*Formenti et al.*, 2002). The extinction profiles derived from LIDAR and SAGE-II are in close agreement with respect to trend and magnitude. The LIDAR profile has been terminated above 10 km due to thick cloud passage. One is able to observe the boundary layer peak at ~2.5 km which is described earlier, as an important parameter for atmosphere mixing (including pollutants). The presence of a cloud results in a sharp enhancement in the extinction and backscatter co-efficient to a high value making the detection quite unambiguous. A small difference in the observed magnitude might due to employed different techniques between LIDAR and satellite, time of observation, mean satellite profile versus a single day LIDAR measurement. The above mentioned height profile of aerosol extinction coefficients obtained using the LIDAR and SAGE-II satellite data are integrated appropriately to obtain the aerosol optical depth (AOD). Generally, we considered the LIDAR profile for the lower height region with respect to the SNR and at higher altitudes from the SAGE-II data. We found the value for February months is around ~0.264 which is in good agreement with AOD measured by the photometer over Johannesburg (0.2966±0.06668).

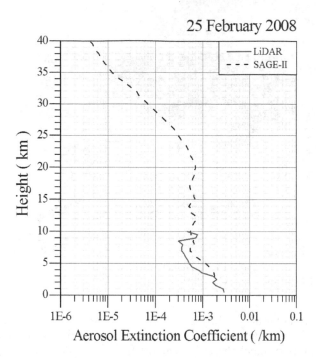

Fig. 15. Height profile of aerosol extinction coefficient derived from LIDAR for the night of 25 February 2008, superimposed by February monthly mean profile of SAGE-II.

4.4 Future perspectives

Based on our knowledge, there are no multi-channel LIDAR systems employed for atmosphere research in South Africa and African countries. Our goal is to achieve a multi channel LIDAR system to address aerosol/cloud, water vapour, lower atmosphere temperature and ozone measurements. LIDAR studies on particulate matter (0.5 and 0.3 microns) elucidate their distribution and concentration in the atmosphere. Particulate matter plays a key role in atmospheric physical and chemical processes from local to global scale. The complexity of these processes have been largely reviewed in literature and LIDAR measurements have mostly contributed to better understanding the role of atmosphere dynamics and particle microphysics. By making observations on a pre-determined spatial scale (from sites to regions) may plausible to calculate atmospheric mass transport and through trajectory analysis to back-track the location of plume sources, e.g. biomass burning. The atmospheric backscatter measurements of aerosols can be used to identify the stratification of pollutants and will enable the classification of the source regions, such as industrial, biological and anthropogenic sources. Later, the plan is to upgrade the system to measure water vapour concentrations in the atmosphere and its localized variations in the lower troposphere. Water-vapour effects global climate change and global warming both directly (water is a primary green-house gas) and through its impact on ecosystems where vegetation sensitivity plays an important feedback role.

Further the ongoing plan is to employ a 2-D scanner into the present LIDAR system (see. Figure 2) will be implemented in near future using a cable/pulley system and an electric winch to lift and lower the scanner. The integration of the scanner assists us in terms of

- X-Y dimensional mapping of the atmosphere (horizontal or vertical cross-section)
- Focusing the target (industrial smoke or cloud of pollutants)
- To study the plume (say smoke, biomass burning and etc), Haze and Aerosol/pollutant dispersion.

Successful implementation of scanner will contribute to LIDAR technology worldwide as, with few exceptions, X-Y dimensional mapping of the atmosphere has not been fully explored. The plan is to include online control of the scanner incorporation of the position of the axes into the present data-acquisition system. The attempt will be done to modify the present data-acquisition software to capture the X-Y cross-sectional display during real time measurements.

5. Acknowledgments

We are thankful to the different South Africa funding agencies addition to the Council for Scientific and Industrial Research-National Laser Centre (CSIR-NLC), Department of Science and Technology (DST), National Research Foundation (NRF) (Grant no: 65086 and 68668), Southern Educational Research Alliance (SERA), African Laser Centre (ALC), Centre National de la Recherché Scientifique (CNRS) (France) and French Embassy in South Africa (France).

6. References

Chiang, C.W. & Nee, J.B. (2006). Boundary layer height by LIDAR aerosol measurements at Chung-Li (25°N, 121°E), *Proceeding of 23rd International Laser RADAR Conference*, 50-6.

Egert, S. & Peri, D. (2008). Automatic retrieval of the atmospheric boundary layer height, *Proceeding of 24th International Laser RADAR Conference*, 320-323.

Fernald, F. G. (1984). Analysis of atmospheric lidar observations – some comments, *Applied Optics, 23*, 652-53.

Fiocco, G. & Smullin, L.D. (1963). Detection of scattering layers in the upper atmosphere (60Å-140 km) by Optical RADAR, Nature, 199, 1275 – 1276.

Fiocco, G. (1984). Lidar systems for aerosol studies, An outline, MAP Handbook, Vol. 13 (ed. R.A. Vincent), pp. 56-58.

Formenti, P; Winkler, H., Fourie, P, Piketh, S., Makgopa, B., Helas, G. & Andreae, M.O. (2002). Aeorosol optical depth over remote semi arid region of South Africa from spectral measurements of the daytime solar extinction and nighttime stellar extinction. *Atmospheric Research, 62*, 11-32.

Hauchecorne, A. & Chanin, M. L. (1980). Density and Temperature Profiles Obtained by Lidar Between 35 and 70 km, Geophys. Res. Lett. 7, 565–568.

Klett J.D. (1981). Stable analytical Inversion solution for processing LIDAR returns. *Appl. Opt.* 20, 211.

Klett, J.D. (1985). LIDAR inversion with variable backscatter to extinction ratios. *Appl. Opt,* 24, 1638-1645.

Ligda, M.G.H.(1963). Proceedings of the first conference on laser technology, U.S. Navy, ONR, 63-72.

5

Smart Station for Data Reception of the Earth Remote Sensing

Mykhaylo Palamar
Department of Devices and Control-Measurement Systems,
Information Technique and Intelligent Systems Research Laboratory
Ternopil National Technical University
Ukraine

1. Introduction

The technology of remote sensing (ERS) provides huge information resources and has the potential to influence the socio-economic development of both security and defence. However, the mass use of remote sensing technologies demands the creation of a network with the technical means of reception and online access to remote sensing data for consumers. The primary source of data for remote sensing is an aerial station (AS), with the reception of information coming from a spacecraft (SC). Typically, these stations are special objects (mainly military), intended to receive, process and disseminate remote sensing data.

For the effective use of ERS data, it is necessary to bring it closer to the end user. This requires universal compact antenna stations of a consumer class, including mobile ones.

This chapter reviews the principles, structures, models and analysis of various technical solutions and the key features, basic functions and control algorithms that are used to create universal automatic ASs (terminals with remote control) and software to control such ASs so as to get remote sensing information from the spacecraft.

The idea of intelligent "personal" aerial station for information receiving is offered proceeding from its function. Such station can be used by small groups or individual researchers directly engaged in contextual information processing i.e. university laboratories, scientific centers, and other organizations interested in such information.

The results of the author's practical experience in creation of remote sensing AS with different types of rotary support devices and with various diameters (from 3 to 12 m) of parabolic reflectors are given. Experimental results of operation of control systems of remote sensing stations using algorithms of artificial neural networks are presented.

2. The structure and principle of the functioning of terrestrial antenna stations for remote sensing data reception

The following conditions are necessary for an ERS system to function:

1. Low-orbital satellites with filming and recording equipment onboard;

2.   Onboard data transmitters via a radio channel;
3.   Terrestrial antenna stations for data reception, its processing and distribution to users.

The general scheme of a satellite monitoring system is shown in Fig. 1.

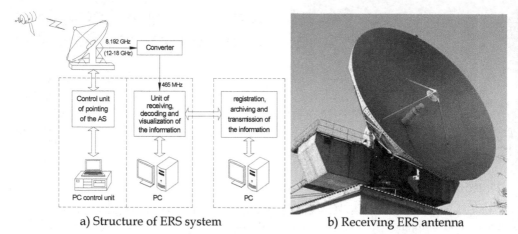

a) Structure of ERS system   b) Receiving ERS antenna

Fig. 1. General scheme of a satellite monitoring system (a). Receiving antenna where the reflector has a diameter of 12 m (b).

According to NASA reports at present 6130 artificial satellites are launched into space. 957 among them are operating on different Earth orbits. Nearly 7 % i.e. more than fifty of them are intended for remote sensing. Nearly 40 countries are directly involved in programmes involving satellite observations and their number is constantly growing. The trend is that the number of spacecraft is growing and the resolution of recharging equipment is increasing (several tens of cm). New technologies of satellite monitoring have appeared (e.g., the miniaturisation of equipment, the usage of micro- and nano-satellites, satellite clusters and the integration of different projects). University (students) satellites and those of other branch research organizations are being launched. New technologies of making survey of necessary territories ordered by customer are applied (Hnatyshyn & Shparyk, 2000).

Ground infrastructure remote sensing systems consist of centres receiving and processing data from spacecraft, with web portals to access the catalogues, archives and operational information from space. The necessary components are: the marketing of software products for thematic data processing systems and the training of qualified personnel.

High-sensitive antenna systems and equipment for reception, demodulation, the decoding of the electromagnetic microwaves from spacecraft and the allocation of the data streams that are encrypted in order to receive data from satellites are all also necessary.

The technology of ERS data reception is more difficult than data reception from geostationary satellites due to the need for tracking remote sensing spacecraft.

Antenna systems with hardware and software controls should automatically direct the focal axis of the reflector of the antenna system into a predictable location point for the spacecraft so as to ensure its tracking. The signals from the satellites are received by the antenna during the spacecraft's tracking.

The structure of the remote sensing antenna complex includes the following main blocks:

- A supporting-rotating device with a pointing mechanism;
- A reflector system mounted on a rotating part of the mechanism;
- A control system for pointing and tracking;
- A system for the receiving, decoding and visualisation of information;
- A system of registration, processing, archiving and the transmission of data.

Satellite trajectories - which are calculated for the next session - are loaded into the PC control unit in a table view before the session with the spacecraft. The control data includes codes for the antenna's angular position and velocity codes for the change. They are transferred to the high-level equipment of the antenna control system from the PC via a communication interface. The PC monitors the antenna position by broadcasting the angular coordinates received from the respective antenna sensors. Moreover, it is necessary to monitor the status of limit switches, the track time, speed and other parameters. The control system needs to be synchronised with a GPS time system in order to ensure the management of the antenna system in real-time.

Information is transmitted via the communication network to the computer after the session's end. The computer has to perform zero-level processing (unpacking the flow and binding the onboard time to terrestrial time) and referencing to geographical coordinates.

2.1 The concept of smart "personal" earth stations for remote sensing

As was noted in the studies of the India Space Department, more often than not remote sensing technology has not yet been effectively used, despite the whole complex of remote sensing satellites available for the country.

The main causes of this are the isolation of consumers from the remote sensing data processing centre, the lack of remote sensing receiving stations and the difficulties involved in gaining access to RS data. Moreover, the important factors are: an insufficient amount of software products and qualified staff in the field of contextual RS data processing, though partially this is a consequence of the reasons already addressed.

Currently, a mainly centralised access method for remote sensing information is used. This approach involves the receiving, processing and dissemination of data only through big centres for space information receiving, often involving military organisations. Such data centres can be compared with the big computer centres from the 1970s that acted as service providers for complex calculations on request. These were non-dynamic structures and ineffective for a wide range of customers. As such, the genuine active development and implementation of informational technologies into daily life began with the popularisation of personal computers, when a wide range of interested consumers became involved in working with information.

However, other technologies related to the distributed method of reception and processing of information are emerging. This information is received locally by organizations interested in such information by means of their own aerial terminals. In such cases the information reaches the user more quickly and more users can work on data processing and analysis concerning their subject-matter. To use this technology, it is necessary to provide users with inexpensive and easy-to-use 'personal' RS data receiving stations. Such stations can significantly change their activities in relation to a number of areas connected to the use of space informational technologies, as with the appearance of the PC.

Personal RS data receiving station is relatively cheap, automated, simple in use (including mobile version) host antenna station designed for use by groups directly engaged in concerned with their subject-matter data processing and decision making (or guidelines in decision-making for management departments). These may be universities, research laboratories, institutes or departments in control organisations. The key characteristic features of such stations should be:

- Compactness and simplicity of operation and maintenance;
- Integration with processing technologies and the storage and thematic analysis of data;
- The use of standard PC configurations;
- Affordable price.

Personal stations allow for the reduction of the access time to remote sensing data and the cheapening and loosening of access for a wide range of users. This solves one of the main requirements of remote sensing data – the efficiency of the acquisition of actual space information about the earth's surface and its objects.

Connecting a wider range of consumers - including the involvement of university science departments and the practical training of staff in the area of thematic data processing - allows for the more effective usage of the satellite in monitoring data for the stable growth and security of countries (according to the GEOSS and GMES programmes, etc.).

The availability of such systems will make remote sensing data an effective information tool for accessing situations and decision-making.

Important features of a personal remote sensing data receiving antenna station should include:

1. The prediction and calculation of the trajectory of spacecraft which are selected by their orbital data from the spacecraft catalogues and the coordinates of the station;
2. Software calibration and the accompaniment of the selected spacecraft on its trajectory with the minimal acceptable error;
3. The tracking of the signal maximum from the spacecraft during its accompaniment and correction of the calculated accompaniment trajectory if necessary;
4. The reception and demodulation of radio-signal selection of the information flow;
5. Real-time data processing;
6. Data visualisation, archiving and storage;
7. Self-checking and the self-diagnosis of the units and the station as a whole;
8. Adaptiveness to the effects of various factors, both external and internal;
9. Connectivity with other stations and external terminals for synchronisation and coordination.

Such functionality would allow the staff to focus on online access and contextual information processing instead of focusing on hardware.

Further, technical problems we had to solve while creating the series of antenna stations for satellite tracking and receiving of remote sensing data as well as broadcasting command information to satellite are described.

2.2 Features and problems that must be addressed during the station's creation

Since the position of the spacecraft for low-orbit remote sensing changes all the time, both hardware and software tools for the controlling and tracking of a satellite in its orbit play an important role in the structure of terrestrial receivers. The required accuracy and acceptable errors in coordinate tracking depends on the chart direction of the aerial and the diameter of its mirror.

Problems involved in AS creation for tracking the remote sensing satellite are caused by the following factors: the low-orbital trajectory of the remote sensing satellite requires the use of a high-dynamic supporting-rotating device for the antenna with the relevant control systems. Increase of image dimensional resolution from the satellite requires the acceleration of the information flow transmission rate which in its turn leads to the enlargement of the reflecting surface diameter of antenna reflector (diameters varying from 3m up to 12m) and its weight as well (Garbuk & Gershenson, 1997).

The speed of information flow is defined as:

$$C = \frac{L \cdot V}{r^2} \cdot I \cdot N \cdot K , \tag{1}$$

where:

L – the width of the Earth's view;
V – the velocity of the sub-satellite point;
I – the number of bites per pixel of the image;
N – the amount of information channels;
K – the coefficient of the coding noise immunity type;
r – the resolution of the Earth's surface survey capability:

$$r \cong \frac{\lambda}{D} \cdot H , \tag{2}$$

where:

λ – the wavelength;
H – the height of the spacecraft;
D – the diameter of the lens.

The larger the diameter of the reflector, the narrower antenna direction chart becomes, which leads to the need to increase dynamic pointing accuracy. For instance, for the AS TNA-57 used for receiving data from the remote sensing Ukrainian satellite 'Sich-2' in the Centre for Space Information Monitoring and Navigation Field Control (CSIM and NFC), the diameter of the antenna reflector is 12 m, its weight is 5,500 kg, while the total weight of the AS is close to 70,000 kg (Fig.1,b). The width chart of the antenna orientation on the level of the 3 dB level is equal to 14 arcmin. Thus, it is necessary to provide speeds of up to 10 degrees / sec with a dynamic tracking error of not more than 1.5 arcmin.

The provision of a large dynamic range of motion for large antennas (a reflector with a diameter of 3m to 12m) and the need to ensure a small dynamic error for spacecraft guidance and tracking are contradicting requirements. Thus, this leads to a more complicated structure and management system for the AS, which increases the cost of the station.

In addition, for classical azimuth-elevation supporting-rotating devices (Fig.1b) there are "dead" zones for spacecraft tracking, for those trajectories that are close to the zenith relative to the location of the terrestrial stations (Belyanstyi & Sergeev, 1980).

3. Structure and algorithms for new constructions of ERS stations

This section discusses some variants of the construction and algorithms of station control systems – as designed by ourselves - which solve the above mentioned problems in order to create effective stations for receiving information from remote sensing spacecraft. The experimental results of their work are given.

3.1 Principles for the functioning of an AS with 3 axes pointing without 'dead zones' accompanying the spacecraft through the zenith

To reduce the high speeds of ASs and to avoid signal loss in the "dead zones" we developed an AS with a 3-axes Support-Rotating Device (SRD) with an implemented additional azimuth axis of E1 with a slope $\gamma \cong 15°$ relative to the direct azimuth axis E3 and a rotation range in the horizontal plane the same as the basic azimuth axis $\pm 170°$ (Fig.2a).

a) antenna "EgyptSat-1"   b) simulation model "EgyptSat-1"

Fig. 2. An AS with a 3-axial SRD (a) and a simulation model of spacecraft accompaniment through the zenith (b).

The aerial control system should perform an orientation of the chart direction of the reflector towards the spacecraft in real-time according to the rule about the spacecraft's motion towards the AS's coordinates. As the basis for the calculation of the orbital motion of the spacecraft, a Keplerian model of the point motion around the static attracting object is accepted. The satellite trajectory is described through Keplerian orbit elements (Fig.3), where:

i – the inclination of the orbiting satellite;

Ω – the longitude of the ascending node from Greenwich during the moment of the epochal time moment T;

ω – the angular distance of the perigee from the ascending node;

p – the orbit parameter dependent on the large semiaxis a: $p=a*(1-e^2)$);

e – orbit eccentricity;

T – epochal time (or time moment). The satellite passes through the point of the ascending node (the intersection of the equator when moving from south to north).

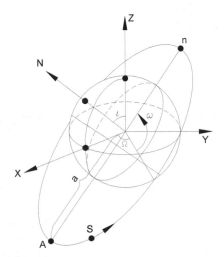

Fig. 3. Parameters of Satellite orbits

However, in reality the movement of the spacecraft is affected by a series of disturbing factors, the most significant of them being: a perturbation of the gravitational anomalies of the Earth, the effect of friction in the upper atmosphere, the influence of the gravity of the Sun and the Moon and the pressure of sunlight. The equation of the spacecraft's motion is described by the system through six differential equations of the first-order with consideration of varying factors. The task of forecasting the spacecraft's movement at every moment of time is reduced to the numerical integration of differential equations of the sixth-order with initial conditions at a given time t_0 (Reshetnev at al., 1988).

Continuously updated data on the spacecraft's orbital parameters is presented in a two-line format (*. TLE) since the calculation of the trajectory can be obtained from the informational satellite catalogues, for instance, on-site http://celestrak.com/NORAD.

The control system calculates the trajectory according to the orbital parameters data in a topocentric coordinate system in an aimer table view $R[t_j, \alpha_j, \beta_j]$, where α_j, β_j – the azimuth angle and the angle of the beam pointing direction of the aerial on the spacecraft at a time t_j.

The control system needs to perform the transformation of input coordinates α_j, β_j in order to accompany the spacecraft with this antenna, from a topocentric azimuth-elevation coordinate system into the local coordinate system of each axis of the AS (array $R[t_j, \alpha 1_j, \alpha 2_j, \alpha 3_j]$), where $\alpha 1_j$, $\alpha 2_j$, $\alpha 3_j$ - the rotation angles of each axis E1, E2, E3 from ERD at time t_j.

To target the spacecraft, the control system controller performs a coordinate conversion according to the algorithm:

$$\alpha 2 = arctg\left(\frac{\cos\gamma \cdot \sin\beta - \sin\gamma \cdot \cos\beta \cdot \cos(\alpha - \alpha 3)}{\sqrt{1 - (\cos\gamma \cdot \sin\beta - \cos\beta \cdot \cos(\alpha - \alpha 3) \cdot \sin\gamma)^2}} \right) + \gamma \qquad (3)$$

$$\alpha 1 = \begin{cases} \alpha_1', \text{ if } X_A \geq 0; \\ \alpha_1' + 180°, \text{ if } X_A < 0 \text{ and } Z_A \geq 0; \\ \alpha_1' - 180°, \text{ if } X_A < 0 \text{ and } Z_A < 0; \end{cases} \qquad (4)$$

where:

$$\alpha_1' = arctg\left(\frac{\cos\beta \cdot \sin(\alpha - \alpha 3)}{\cos\gamma \cdot \cos\beta \cdot \cos(\alpha - \alpha 3) + \sin\gamma \cdot \sin\beta} \right) \qquad (5)$$

$$X_A = \cos\gamma \cdot \cos\alpha 3 \cdot \cos\alpha \cdot \cos\beta + \sin\gamma \cdot \sin\beta + \cos\gamma \cdot \sin\alpha 3 \cdot \cos\beta \cdot \sin\alpha ,$$

$$Y_A = -\sin\gamma \cdot \cos\alpha 3 \cdot \cos\beta \cdot \cos\alpha + \cos\gamma \cdot \sin\beta - \sin\gamma \cdot \sin\alpha 3 \cdot \cos\beta \cdot \sin\alpha ,$$

$$Z_A = -\sin\alpha 3 \cdot \cos\beta \cdot \cos\alpha + \cos\alpha 3 \cdot \cos\beta \cdot \sin\alpha ,$$

$\alpha 1$ – the rotation angle of the main azimuth at axis E1,
$\alpha 2$ – the rotation angle of the elevation axis E2, and
$\alpha 3$ – the rotation angle of the azimuth at vertical axis E3.
$\gamma \cong 15°$ - the angle of the axis E1 relative to the axis of E3.

The range of angle changes:

α - $(0 \div 360°)$,
β - $(0 \div 90°)$,
$\alpha 1, \alpha 3$ - $(0 \div \pm 170°)$,
$\alpha 2$ - $(0 \div 120°)$.

During the execution of the accompaniment of a spacecraft with a given aimer table (array $R[t_j, \alpha_j, \beta_j]$), the controller control system has to convert them into a format of local coordinates (array $R[t_j, \alpha 1_j, \alpha 2_j, \alpha 3_j]$).

To determine the real data about the AS's position and to compare with a given aimer table and issue them in the control and information processing centre, it is necessary that the inverse transformation of the "local" coordinate axes in the system topocentric coordinates pointing to the spacecraft accord with the correspondences below:

$$\alpha = \begin{cases} \alpha', \text{ if } X_B \geq 0, Z_B \geq 0; \\ \alpha' + 360°, \text{ if } X_B \geq 0 \text{ and } Z_B < 0; \\ \alpha' + 180°, \text{ if } X_B < 0; \end{cases} \qquad (6)$$

Where:

$$\alpha' = arctg\left(\frac{\cos\gamma\cdot\sin\alpha3\cdot\cos(\alpha2-\gamma)\cdot\cos\alpha1-\sin\gamma\cdot\sin\alpha3\cdot\sin(\alpha2-\gamma)+\cos\alpha3\cdot\cos(\alpha2-\gamma)\cdot\sin\alpha1}{\cos\gamma\cdot\cos\alpha3\cdot\cos(\alpha2-\gamma)\cdot\cos\alpha1-\sin\gamma\cdot\cos\alpha3\cdot\sin(\alpha2-\gamma)-\sin\alpha3\cdot\cos(\alpha2-\gamma)\cdot\sin\alpha1}\right)$$

$$X_B = \cos\gamma\cdot\cos\alpha3\cdot\cos(\alpha2-\gamma)\cdot\cos\alpha1 - \sin\gamma\cdot\cos\alpha3\cdot\sin(\alpha2-\gamma) - \sin\alpha3\cdot\cos(\alpha2-\gamma)\cdot\sin\alpha1;$$

$$Y_B = \sin\gamma\cdot\cos(\alpha2-\gamma)\cdot\cos\alpha1 + \cos\gamma\cdot\sin(\alpha2-\gamma); \tag{7}$$

$$Z_B = \cos\gamma\cdot\sin\alpha3\cdot\cos(\alpha2-\gamma)\cdot\cos\alpha1 - \sin\gamma\cdot\sin\alpha3\cdot\sin(\alpha2-\gamma) + \cos\alpha3\cdot\cos(\alpha2-\gamma)\cdot\sin\alpha1.$$

$$\beta = arctg\left(\frac{\cos\gamma\cdot\sin(\alpha2-\gamma)+\sin\gamma\cdot\cos(\alpha2-\gamma)\cdot\cos\alpha1}{\sqrt{1-(\sin\gamma\cdot\cos(\alpha2-\gamma)\cdot\cos\alpha1+\cos\gamma\cdot\sin(\alpha2-\gamma))^2}}\right) \tag{8}$$

The control system of such an AS needs to calculate and execute the required angle $\alpha3$ vertical azimuth axis E3 after every calculation or after receiving - via the communication channel - the trajectory of spacecraft, taking into account the mechanical limits of the rotation range of this axis, as follows:

$$\alpha3 = \alpha_M , \text{ if } 0 \le \alpha_M \le \alpha_{\theta^+};$$

$$\alpha3 = \alpha_{\theta^+} , \text{ if } \alpha_{\theta^+} < \alpha_M \le 180°;$$

$$\alpha3 = \alpha_{\theta^-} , \text{ if } 180° < \alpha_M < 190°;$$

$$\alpha3 = \alpha_M - 360° , \text{ if } 360° + \alpha_{\theta^-} \le \alpha_M \le 360°;$$

where:

$\alpha_{\theta^+}, \alpha_{\theta^-}$ - the angles of triggering the limit switches the constraint turn of the antenna on the angle $\alpha3$ (around an axis E3) into "plus" and "minus" respectively ($\alpha_{\theta^+} \approx 170°$; $\alpha_{\theta^-} \approx -170°$);

α_M - a value of azimuth counting with a maximum angle of the elevation of the spacecraft ($\alpha_M = \alpha(t)$ at $\beta(t) = \beta_{max}$), determined from the pointing-table that is calculated for the selected spacecraft.

The calculation of the angles $\alpha1(t)$ i $\alpha2(t)$ is performed by the use of angles $\alpha(t)$, $\beta(t)$ and $\alpha3$.. Such an AS design and algorithm are implemented in the terrestrial bilateral An AS to manage and control the spacecraft telemetry RS «EgyptSat-1" is installed and operated in Egypt (Fig. 2a).

Fig.4a shows the diagram of "Terra" spacecraft's tracking trajectory through zenith (the maximum lifting angle = 90°) in the system of azimuth-elevation coordinates **R** [t, a, β] of topocentric coordinate system. The crimson diagram represents targeted angles on azimuth and the yellow one - on angular altitude. Fig.4b represents diagrams of tracking after

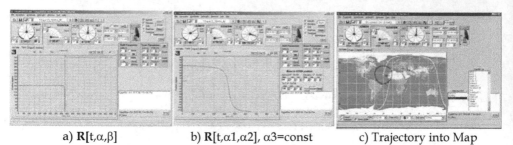

a) **R**[t,α,β]   b) **R**[t,α1,α2], α3=const   c) Trajectory into Map

Fig. 4. Graphs of the trajectory of spacecraft tracking via the zenith: (a)- for a 2-axis AS in topocentric coordinates **R**[t,α,β]; b- for a local axis E1, E2, **R**[t,α1,α2,α3]. The axis E3 is fixed at 107° during the session.

trajectory conversion from topocentric coordinate system into coordinate system of antenna axes **R** [t, α1, α2, α3]. The bottom straight azimuth axis E3 before the beginning of session for the given trajectory is to rotate the antenna system on azimuth towards the direction of the maximum elevation of the spacecraft for chosen trajectory constituting in this case the angle 106° 30 min.

As can be seen from the graphs, at a spacecraft's zenith point a velocity of an azimuth axis for a classic 2-axial AS tends towards infinity (Fig.4a). After the conversion to a 3-axis coordinate system (Fig.4b), the maximal accompaniment speed of the inclined azimuth axis is not more than 2.5 degree / sec. This enables the reduction of dynamic errors during the tracking of the spacecraft.

With the exception of the software method for tracking on a pre-calculated trajectory of the spacecraft, the AS control system implements the tracking of the spacecraft by an auto-tracking method of a signal finder with the goal of supporting a maximum value of the signal. It is also possible to use a compound method of software tracking with automatic correction of tracking table according to the signal and additional manual control.

Total-difference (monoimpulse) type of aerial-feeder device (Fig.5) is used in the designed aerial system for the execusion of satellite automatic tracking according to direction finder signal. Besides the main total informational signal the difference signals on each coordinate forming aerial direction finder characteristic are received on its output. The differencing signal provides information about the value and the sign of an error deviation of the AS from the signal maximum.

Fig.6 shows a graph of the error of the antenna beam's angular deviation from the desired trajectory in angular minutes (over the time t = 220 s) which is not exceeding - as seen from the graphs - 4 angular minutes.

In general, the total combined error of the tracking is a function of time and depends upon the parameters of the control system and the characteristics of the controlling and disturbance signals that affect the system during the process of tracking the spacecraft. As such, the maximum efficiency of remote sensing information reception is achieved with the minimum total tracking error.

Subsection 4 is devoted to a search for the structures and algorithms for efficient system operation employing the use of artificial neural networks.

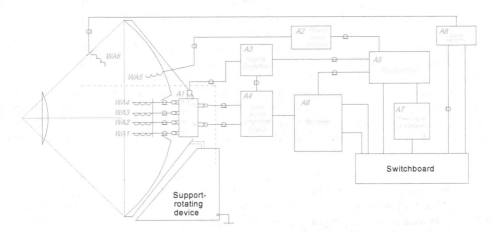

Fig. 5. Block-scheme of an antenna-feeder device of a total-difference (mono-impulse) type.

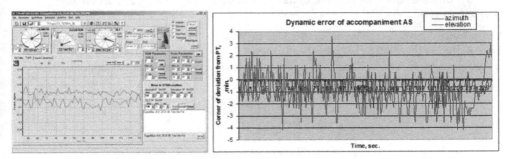

Fig. 6. Graph of the error of the antenna beam's angular deviation from the desired trajectory in angular minutes (over the time t=220 s).

Due to the enhancement of AS design and control algorithms, the speed of moving object tracking in the culminating moment of the spacecraft is significantly reduced, which reduces the requirements for the electromechanical components of the AS and allows the reduction of the dynamic errors involved in tracking. Structural and algorithmic solutions are implemented and tested in AU "Egyptsat-1".

3.2 Antenna System with a rotary device based on the six-axis Stewart platform (Hexapod scheme)

The disadvantages of all types of classic two-axial and modified three-axial SRD constructions of ASs involve their complexity and the high requirements for the accuracy of rotating mechanisms with a large diameter. This makes antenna systems too ponderous, their support-rotating devices too complex for manufacturing and assembling, and their cost too expensive.

Recently, for tracking along complicated trajectories mechanisms of manipulators with parallel kinematic units especially based on six-axis Steward platform (Fig.7) are widely used in robotics, machine-tool constructions, benches and other equipment (Stewart, 1965;

Fichter, 1986). Such mechanical systems consist of platforms connected by a system of variable (controlled) length sections, and they have a certain advantages over rotary mechanisms. For example, a combination of hardness and compactness, reliability, ease of design, manufacturability and studies (Nair & Maddocks, 1994; Kolovsky at al., 2000; Afonin at al., 2001). The Stewart platform is the subject of many scientific studies. There are examples of their use in some of the application problems provided by the data from the booklets of companies and technical exhibits, but the use of parallel kinematic mechanisms based on the Stewart platform in the mechanisms of the SRD of ASs for tracking various spacecraft trajectories - including low-orbital remote sensing satellites - has not yet been investigated.

Below we consider the construction and imitation of a model of the AS support-rotating device based on the six-degree Stewart platform (Hexapod scheme) as an alternative to traditional support-rotating devices. We investigated the possibilities and features of such an AS in performing the tracking of low-orbital satellites.

3.2.1 Specifics of the schema and construction of an AS with a support-rotating device Hexapod

A support-rotating device based on a linear drive (Fig.7) consists of two platforms, one of them is the basis of SRD and the other is the basis for binding the reflector of the satellite and six actuators, each attached to the upper and lower platform via a cardan joint.

Fig. 7. Six-axis Stewart platform.

In our laboratory, we developed a research model for the construction of an AS with a support-rotating device based on the Stewart platform (Hexapod) and a control system for it (Fig.8).

The carcass of this support-rotating mechanism has six points of freedom which allows it rotate the reflector in the air with high accuracy.

A support-rotating device of this construction has benefits comparative with classic rotary mechanisms:

- Simplicity of mechanical construction, toughness, easy access to mechanical units of aerial, absence of cable twisting;
- No "dead" zones during satellite tracking;

Fig. 8. Antenna System with a support-rotating device based on the Stewart platform (Hexapod).

- No restrictions on rotation on the azimuth axis;
- The low speed of driving actuators for any tracking trajectories of a satellite;
- High accuracy in aiming;
- The ability to work in difficult conditions;
- Relatively low cost.

The main disadvantages of this type of support-rotating device include some limitations at low tilt angles of the reflector and the complexity of the simultaneous motion control of six actuators. Unlike classical AS support-rotating devices, the control of the support-rotating device based on a linear drive demands the precise coordination of the parallel movement of all six actuators simultaneously. The closing of every actuator must always lie in corresponding areas, otherwise the construction may be destroyed or the actuators may fail.

3.2.2 Algorithm to control AS based on linear circulating platform

In common case to point the aerial beam on the given azimuth and location angle it is necessary to set the lengthening of each actuator on certain value. In order to find the motion laws of actuators let us solve the inverse problem.

Let us define a plane of the support-rotating device in a Cartesian coordinates system with x, y, z, axes to which the reflector of the antenna is mounted. Since the physical size of the upper platform and the mount points of the actuators on it are known, it is possible to find the coordinates of the hinges. Similarly, let us set the base of the support-rotating device (the lower platform) basis and determine the coordinates of the lower hinges.

At the maximal lengthening of the actuators, the planes will be maximally remote from each other. At the minimum lengthening the distance between them, it will be at the minimum (Fig. 9). In extreme positions, the planes can be located only when parallel to each other.

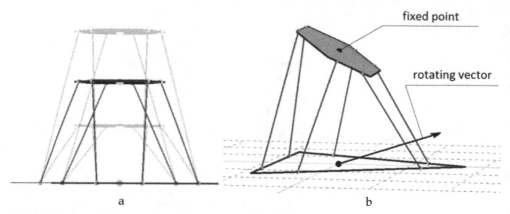

Fig. 9. Location of the platform plane at the different lengthening of actuators.

It is clear that the upper platform has to be in the middle position in order to achieve the maximal possible turn of the antenna reflector. As such, the equal motion of the actuator is kept both upwards and downwards.

Let us perform a turn of upper plane with the hinges mounted accordingly into it, making use of affine isometric transformations of the coordinates.

Three parameters are needed to perform the arbitrary rotation in space:

- A fixed point of transformation;
- A vector that is the centre of the rotation;
- A rotation angle value φ.

Let us choose a point in the centre of the upper platform as a fixed point that passes into itself (as a result of rotation) (Fig.9b). Consider a vector (i.e., the centre of rotation) set by two points p1 and p2:

$$\mathbf{v} = p2 - p1 \tag{9}$$

The direction is determined by the order of using these points. Only the direction of this vector is important. Its position in space does not affect the rotation result.

Let us perform a rotation axis vector normalisation to simplify the operation's execution: replace it with the vector of unit length. The second vector has the same direction in space as the first one:

$$
\begin{aligned}
S &= \sqrt{X^2 + Y^2 + Z^2} \\
X_N &= X/S \\
Y_N &= Y/S \\
Z_N &= Z/S
\end{aligned}
\tag{10}
$$

The rotation is partly simplified if the fixed point (together with the rotation object) is in the zero point of the coordinates. Thus, the first operation of transformation is T(-p0), and the last is T(p0). Where T(-p0) and T(p0) are the appropriate matrices of transformation (Shikin & Boreskov, 1995):

$$T(P_0) = \begin{bmatrix} 1 & 0 & 0 & \alpha_x \\ 0 & 1 & 0 & \alpha_y \\ 0 & 0 & 1 & \alpha_z \\ 0 & 0 & 0 & 1 \end{bmatrix}$$ (11)

$$T(-P_0) = \begin{bmatrix} 1 & 0 & 0 & -\alpha_x \\ 0 & 1 & 0 & -\alpha_y \\ 0 & 0 & 1 & -\alpha_z \\ 0 & 0 & 0 & 1 \end{bmatrix}$$ (12)

Thus, the matrix of a complex transformation will have such a form:

Rotation around an arbitrary axis reduces in relation to the consequent rotation around the particular coordinate axes. The main problem is to find the rotation angles for every axis.

Let us execute the first two rotation operations to combine the rotation axis **v** with the coordinate axis Z. Next, rotate the object around the axis Z to a necessary angle and execute the previous two turns in reverse order.

Accordingly, the matrix of the complex transformation has the form:

$$M = R_x(-\theta_x)R_y(-\theta_y)R_z(\theta_z)R_y(\theta_y)R_x(\theta_x)$$ (13)

The determination of the matrices $R_y(\theta_y)$ and $R_x(\theta_x)$ form the most difficult part of the calculations.

We will consider the components of vector **v**. As **v** is the vector of unit length, then:

$$a_x^2 + a_y^2 + a_z^2 = 1$$ (14)

Let us draw a segment from the beginning of the coordinates to the point (a_x, a_y, a_z). This segment will have a unit length and the same direction as the vector **v**. Drop the perpendiculars from a point (a_x, a_y, a_z) to every coordinate axis as it is represented by Fig. 10. Three direction angles - φ_x, φ_y, φ_z - are the angles between the vector **v** and the coordinate axes. The correlation between direction cosines and the components of vector **v** are:

$$cos\ \varphi_x = a_x$$

$$cos\ \varphi_y = a_y$$ (15)

$$cos\ \varphi_z = a_z$$

Only two direction angles are independent, because:

$$Cos^2\varphi_x + Cos^2\varphi_y + Cos^2\varphi_z = 1$$ (16)

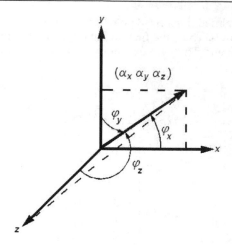

Fig. 10. Direction angles of elevation.

Knowing the values of the direction cosines, it is possible to calculate the value of the Θx and Θy angles. As we see in Fig.11, the rotation of point (a_x, a_y, a_z) will lead to the segment rotation where it will be located on the plane y=0. The length of the segment projection (before the turn) on the plane x=0 is equal to d.

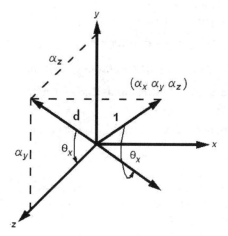

Fig. 11. Rotation angle placement according to the X axis.

Since the rotation matrix contains sines and cosines instead of angles, there is no need to find the Θx value itself, so the rotation matrix $Rx(\theta x)$ will be:

$$R_x(\Theta_x) = \begin{bmatrix} 1 & 0 & 0 & 0 \\ 0 & \cos\theta & -\sin\theta & 0 \\ 0 & \sin\theta & \cos\theta & 0 \\ 0 & 0 & 0 & 1 \end{bmatrix} = \begin{bmatrix} 1 & 0 & 0 & 0 \\ 0 & \alpha_z/d & -\alpha_y/d & 0 \\ 0 & \alpha_y/d & \alpha_z/d & 0 \\ 0 & 0 & 0 & 1 \end{bmatrix} \tag{17}$$

And the inversed rotation matrix Rx(-θx) will be:

$$R_x(-\Theta_x) = \begin{bmatrix} 1 & 0 & 0 & 0 \\ 0 & \cos\theta & \sin\theta & 0 \\ 0 & -\sin\theta & \cos\theta & 0 \\ 0 & 0 & 0 & 1 \end{bmatrix} = \begin{bmatrix} 1 & 0 & 0 & 0 \\ 0 & \alpha_z/d & \alpha_y/d & 0 \\ 0 & -\alpha_y/d & \alpha_z/d & 0 \\ 0 & 0 & 0 & 1 \end{bmatrix} \quad (18)$$

The elements of the $R_y(\theta_y)$ matrix are calculated in a similar way (Fig.12).

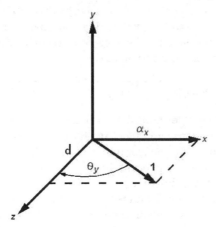

Fig. 12. Rotation angle placement according to the Y axis.

The corresponding rotation matrices are:

$$R_y(\Theta_y) = \begin{bmatrix} \cos\theta & 0 & \sin\theta & 0 \\ 0 & 1 & 0 & 0 \\ -\sin\theta & 0 & \cos\theta & 0 \\ 0 & 0 & 0 & 1 \end{bmatrix} = \begin{bmatrix} d & 0 & -\alpha_x & 0 \\ 0 & 1 & 0 & 0 \\ \alpha_x & 0 & d & 0 \\ 0 & 0 & 0 & 1 \end{bmatrix} \quad (19)$$

$$R_y(-\Theta_y) = \begin{bmatrix} \cos\theta & 0 & -\sin\theta & 0 \\ 0 & 1 & 0 & 0 \\ \sin\theta & 0 & \cos\theta & 0 \\ 0 & 0 & 0 & 1 \end{bmatrix} = \begin{bmatrix} d & 0 & \alpha_x & 0 \\ 0 & 1 & 0 & 0 \\ -\alpha_x & 0 & d & 0 \\ 0 & 0 & 0 & 1 \end{bmatrix} \quad (20)$$

Thus the rotation axis (vector **v**) coincided with the axis Z. Then let us perform the rotation on needed angle elevation (angle of the aerial reflector beam pointing):

$$R_z(\Theta_z) = \begin{bmatrix} \cos(\Theta) & -\sin(\Theta) & 0 & 0 \\ \sin(\Theta) & \cos(\Theta) & 0 & 0 \\ 0 & 0 & 1 & 0 \\ 0 & 0 & 0 & 1 \end{bmatrix} \quad (21)$$

After that we carry out the reverse transformations: $R_y(-\theta x)$, $R_x(-\theta y)$, $T(-p0)$ and obtain the top plane rotated on the given pointing angle corresponding with the aerial beam elevation angle. As a result of the multiplication of all the discovered transformation matrices, we will get the complex matrix M:

$$M = T(-p_0)R_x(-\theta_x)R_y(-\theta_y)R_z(\theta_z)R_y(\theta_y)R_x(\theta_x)T(p_0) \tag{22}$$

The multiplication of an arbitrary point in a three-dimensional space on a specified complex matrix will cause it to turn around to some fixed point in the same space.

After the rotation of the upper platform, we receive new coordinates of the upper ends of actuator hinges used to mount to the platform. Having the coordinates of the upper and lower hinges in space, we calculate the distance between them using a correlation (23) (the actuator lengthening that it was necessary to find):

$$S = \sqrt{(x_2 - x_1)^2 + (y_2 - y_1)^2 + (z_2 - z_1)^2} \tag{23}$$

On the basis of the resulting algorithm, the simulation work program is developed. This program provides the calculations and represents the position of the actuators and the rates of their movement depending on the azimuth and elevation angle with the reflection of three-dimensional model of the supporting-turning device and antenna (Fig.13). In this model, it is possible to set the different geometrical parameters of the support-rotating device's construction (Fig.14). Different dimensions of the construction for determination of various optimum correlation between minimum values of the inclination angles, speeds and accuracy of work and control actuator motion while constructing the control system can be set in the model.

The control of a supporting-rotating device of a Hexapod-type requires precise (coordinated in time) cooperation between the position sensors and the delivery system of the control signal for all six drives. It is needed in order to preserve the system's integrity and to avoid

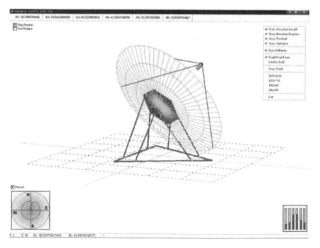

Fig. 13. A three-dimensional model of the support-rotating device.

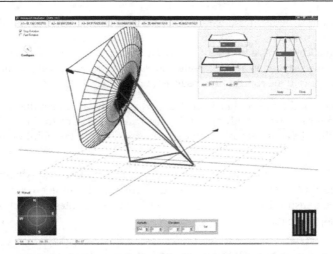

Fig. 14. Modelling of the constructional parameters of the support-rotating device.

physical damage. The six actuators form a single system. The control system must provide the simultaneous coordinated parallel control of 6 drives. The developed control system implements the algorithms of parallel work on the basis of a FPGA programmable logical integrated circuit. The block diagram with the cooperation chart of the control system's basic nodes is represented in fig.15.

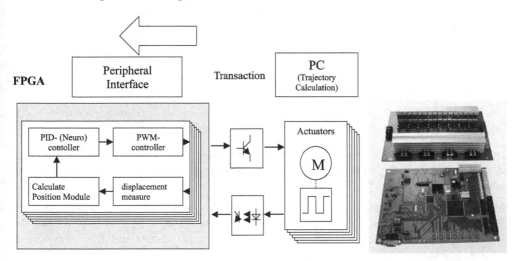

Fig. 15. Interaction scheme between the main units of the AS control system with a Hexapod.

The computer for the control system generates the array of the points for each actuator which create the trajectory. Every point is transacted to an FPGA that has six logical channels generated for it. Every channel is responsible for the work of a corresponding actuator, and consists of a PID regulator, a PWM inspector, a processing module for actuator sensor signals and a calculation module for the actuator's current position. In order to call the resources of every channel, a module is created. It provides an interface to access the periphery and provides its own address space for every channel and ensures the integrity of the data passed. Additionally, an interrupt controller is created so as to increase the reaction of all the system. This controller signals to the control processor regarding emergency events.

All of the channels of the control block work synchronously. This provides for simultaneous data reading from the sensors with processing and control actions for all of the actuators. It provides work for all 6 actuators as a single system for tracking the pointing trajectory of the spacecraft.

The graphs of the aimer table transformations from the topocentric system are shown in Fig.16. A trajectory is set by the arrays of the azimuth and elevation coordinates ($R[t_j, \alpha_j, \beta_j]$). These arrays are transformed into the local movement coordinates for each actuator (array $R[t_j, \alpha 1_j, \alpha 2_j, \alpha 3_j, \alpha 4_j, \alpha 5_j, \alpha 6_j]$).

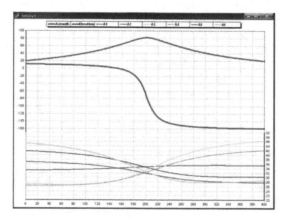

Fig. 16. Graphs of the aimer table transformation in the topocentric coordinate system ($R[t_j, \alpha_j, \beta_j]$) and in the local coordinate system $R[t_j, \alpha 1_j, \alpha 2_j, \alpha 3_j, \alpha 4_j, \alpha 5_j, \alpha 6_j]$.

The control program on the control system computer provides the visualisation of a movement diagram for each actuator and their speed; it also provides the calculation of trajectory tracking errors (fig.17).

So, the supporting-rotating device of an aerial system constructed on the basis of a Stewart platform (parallel kinematics structure Hexapod) considerably simplifies the mechanical construction of the AS, but increases the requirements for the schema and algorithms of the control system.

4. The use of neural network technology in the control systems of ERS aerial stations

The calculations of an AS's dynamic parameters for the construction of apparatus-programming devices for aerial guidance control according to the classical method - especially for six-wheeled or six-drive traversing mechanisms Hexapod - are connected with

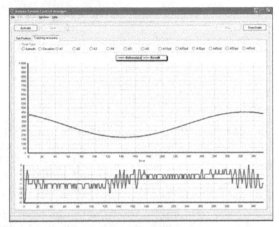

Fig. 17. Graph of the tracking of axis 1 of the actuator.

technical difficulties relating to the determination of the series of the AS's real parameters. These include, modulus inertia moments, changes of resistance friction depending on the inclination angle and the ratio of the aerial modulus position for various axes, the rigidity changes of mechanical transmissions, clearances, the instability of electric drive characteristics, the stochastic influence of wind loadings, the possible instability of time-sampling and program data processing during coordinates transformation, etc. Such mechanical systems essentially have a non-linear character. The methodological maintenance for the control of multidimensional interconnected dynamic units of such mechanical systems has not been solved sufficiently.

4.1 The AS model and its separate elements in the control system

One of the most effective and important methods for the control of dynamic objects with indistinctly determined parameters is the use of an algorithm of a proportional-integral-differential (PID) controller with the adaptive adjustment of PID-coefficients:

$$u(t) = K_p \left[\varphi(t) + \frac{1}{T_I} \int_{T-\Delta t}^{T} \varphi(t)dt + T_D \frac{d\varphi(t)}{dt} \right], \qquad (24)$$

This expression is converted into digital form, convenient for program-realisation on the microcontroller:

$$u(t) = u(t-1) + K_P(e(t) - e(t-1)) + K_I e(t) + K_D(e(t) - 2e(t-1) + e(t-2)) \qquad (25)$$

where u(t) – the regulator output signal;

$\varphi(t)$ – the deflection of angular position from the needed target;
K_p – the amplification factor in the return circuit;
T_I, T_D – the time differentiation and integration constants;
$e(t) = r(t) - y(t)$, - the regulation error;
$r(t), y(t)$ – the target and the value of output signal for the object quidance;
K_P, K_I, K_D – PID coefficients requiring optimal adjustment.

The discrete transfer function of such a controller is determined by the expression:

$$W_p(z) = k_p \left[1 + \frac{T_0(1+z^{-1})}{2T_I(1-z^{-1})} + \frac{T_D}{T_0}(1-z^{-1}) \right]$$ (26)

T_0 - is the quantisation time, able to adjust adaptively depending on the divergence angle while approaching a given coordinate.

However, in dynamic processes with variable parameters and interferences, it is rather difficult to ensure optimal coefficient adjustments. Very often, parameters for adaptive control should be chosen by a method of trial and error. There are a wide range of methods and algorithms for PID-controller self-adjustment, mostly resulting in the complication of algebraic calculations and requiring the introduction of many new system parameters (Kuncevych, 1982).

One of the alternatives to the classical models and methods is the creation of a control model based on the use artificial neural networks (ANNs). ANNs are a group of algorithms described and modelled according to principles analogous to the work of human brain neurons. A neuron network is able to compare its output signal with a given training signal and carry out self-adjustment according to certain criteria by means of the automatic selection of various internal weighting factors aimed at minimising the difference between the actual output signal and the training signal.

The functional characteristics of neuron networks show that this technology can provide control results much better than those obtained by means of classical controls and software (Miroshnik at al., 2000; Callan, 2001). The great value of ANN use lies in its universal solution for various types of control objects distinguished by the different parameters set, i.e., the different electro-mechanical modulus of ASs and the various types of mounting-traversing device structures and loadings (Golovko, 2001; Zaichenko, 2004). ANNs are not programmed but taught, which is why their solution quality depends mainly upon the data quality and the quantity of data needed for teaching.

4.2 Neural network use for the optimisation of control parameters

The idea the use of ANNs in aerial movement control systems is that the main control parameters (PID-coefficients, etc.) are ANN outputs adjusted while working through a series of test orbits of AS movements, i.e., ANN teaching (Omata at al., 2000). The scheme of ANN use in an AS's axes control circuit is shown in Fig.18.

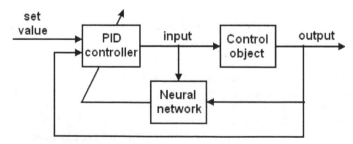

Fig. 18. A scheme for neuron control with self-adjustment.

Fig.19 reflects the structural scheme of a 3-contour AS control system for each of 3 axes of the aerial station "EgyptSat-1" using a neuro-controller for optimal coefficient adjustment in an external control contour.

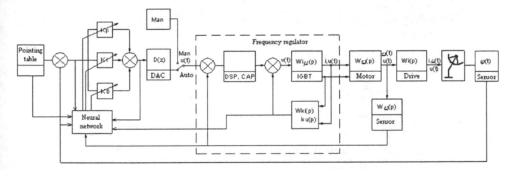

Fig. 19. Structural-algorithmic scheme of an AS control contour with a neuro-controller.

The internal contour is directly closed in the frequency regulator which controls the voltage and the current of the electric drive for local rotation control. The second is the contour of the AS's axes rotation speed control. The external control contour is closed on the angular position of the AS axes.

A model of an AS control system and the submodels of its separate units (aerial, controller, frequency regulator, motor, Fig.20) are constructed due to the program complex MatLab/Simulink.

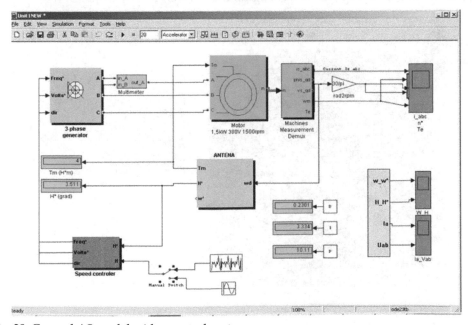

Fig. 20. General AS model with a control system.

The unit for the adjustment and optimisation of the PID-controller's parameters Optimum_1 is introduced into a submodel of the controller Speed controller (Fig.21).

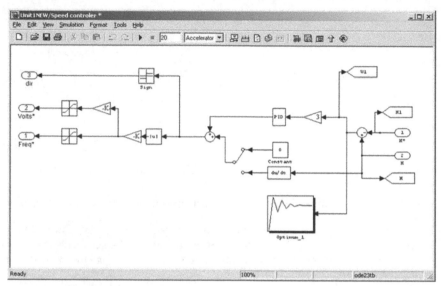

Fig. 21. Model of a guidance controller.

Error limits on AS movement deviations from the test sinusoidal guidance table provided within the limits of 0.2 degrees are set in the optimisation unit Block Parameter (Fig.22).

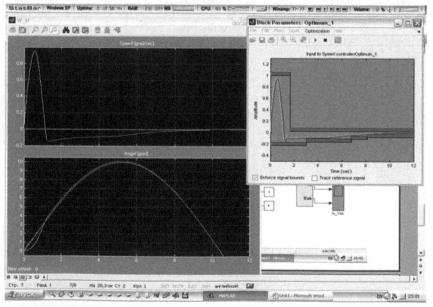

Fig. 22. The process of PID-control coefficients optimisation.

From the previous results in initial sections of GT, we can observe that considerable deviations occur as a result of the dynamic resistance moments during the AS's acceleration. To perform an optimal coefficient adjustment, the error limits are extended on the initial orbit section up to 1.0 degree (Fig.22), otherwise the ANN cannot adjust.

As the result of modelling, the deviation error diagram from the GT can be obtained (Fig.23).

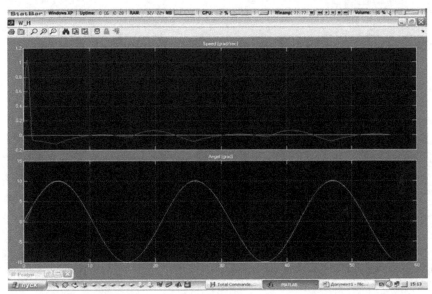

Fig. 23. The modelling of deviation errors for AS tracking along the sinusoidal GT.

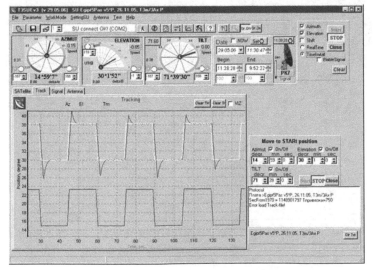

Fig. 24. Adjustments of the rate regulation of the impulse functions on 2 axes ($\alpha 1$, $\alpha 2 = 8°$).

The results of control PID-coefficient adjustment were tested on the 3-axes AS "EgyptSat-1" with the perfecting of various test orbits, and especially generated impulse functions (Fig.24, Fig.26), sinusoidal functions (Fig.25), special "high-speed" tables of target designations (Fig.27) and real satellite orbits.

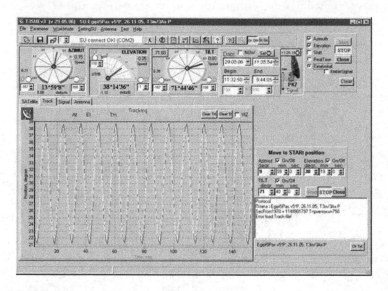

Fig. 25. Adjustments of the rate regulation on sinusoidal functions ($\alpha 2 = 60°$).

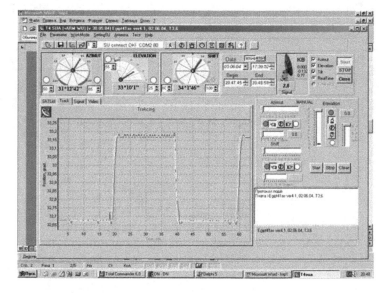

Fig. 26. Diagram of the impulse AS orbit perfection along the β axis.

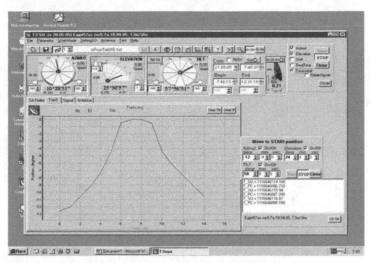

Fig. 27. Test orbit with a maximum tracking speed of 5 degree/sec.

4.3 Neural network use in the contour of aerial axes control

Another structure of neural AS control like dynamic object is offered. In this structure neural network and common PID-controller are used at the same time (Fig.28).

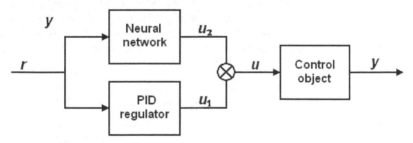

Fig. 28. Parallel scheme of a neuro-controller.

A typical two-layer perceptron with 10 neurons in an intermediate layer was chosen for the contour of the AS's axes control. Synthesis was carried out with the NNTOOL utility and MATLAB MEDIUM. A functional model of the system with a PID and neuro-controller was created with the Simulink program (Fig.29). The neuro-controller emulates the operation of the PID-controller. Neuron network teaching was executed via the method of reverse error extension. For this purpose, a set of teaching pairs - "input vector"/"right output" - were generated. In such a case, the input vector enters the network entrance and the state of all the intermediate neurons is calculated in series, while the output vector is compared with the right one and formed at the exit. Deviation provides errors which extend in the reverse direction along the network connection; afterwards, weighting factors are corrected to rectify it. After repeating this procedure a thousand times, we managed to teach the neuron network.

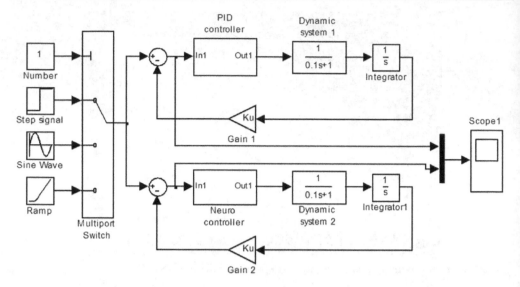

Fig. 29. Functional comparison model of systems with a PDF and a neuro-controller.

Fig.30 depicts the results following the neuron network's operation. Evidently, a simple multilayer perceptron (red colour graphics) had worse results in comparison with the PIF-control. The application of a recurrent perceptron distinguishing from the previous one by presence of delay lines on entries has better results (Fig.31). However, insufficient teaching stability marks its disadvantage. Imitative modelling shows that during the optimal selection of neuron network topology and the teaching of algorithms, it is possible to use it for the effective control of complex dynamic objects, such as large-sized aerial complexes.

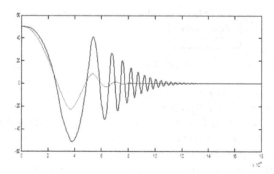

Fig. 30. Comparison of the PIF-controller's operation with a multilayer perceptron.

By introducing the neuron network into the control scheme, it can be used for the more effective operative adjustment of control parameters by means of its teaching of various test orbits. The strategy of neuron control with self-adjustment can be used for different types of AS drives with various dynamic characteristics.

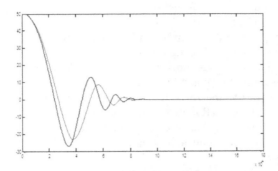

Fig. 31. Comparison of the PDF-controller's operation with a recurrent perceptron.

5. Conclusion

The investigation and search for optimal structures for mounting-traversing devices and control systems for the construction of aerial stations for remote sensing data reception have been carried out in this work. The models and results of the operation of two types of mounting-traversing AS devices have numerous advantages when compared with classical models and can be used for the creation of personal aerial stations for remote sensing data reception, as shown. The application of neuron networks in the control systems of ASs for remote sensing data reception can provide for the more accurate operation of control systems for satellite guidance and their tracking along the orbit in spite of the faults relating to constructional and dynamic AS parameters. The use of a neuron network in a control circuit also provides considerable advantages over traditional control systems due to the fact that for their realisation there is no need for accurate mathematical models of control objects.

6. References

Afonin V.L., and Krainov A.F., Kovalev V.E., Lyakhov D.M., Sleptsov V., Processing equipment of new generation. - Design concept Moscow: Mashinostroenie, 2001, 256 p.

Belyanstyi P.V., Sergeev B.G. Control of terrestrial antennas and radio telescopes. – M.: Sov. Radio, 1980. – 280 c.

Callan R., The basic concept of neural networks. - Moscow: Publishing House "Williams", 2001. - 288.

Fichter E.F., A Stewart platform – based manipulator , general theory and practical construction. - International Journal of Robotics Research. 1986. Vol. 5, No. 2, pp. 157 – 182.

Garbuk S.V., Gershenson V.E. Space remote sensing. – M.: publishing house A and B, 1997. – 296 c.

Golovko V.A., Neural networks: training, organization and application. - M.: IPRZHR, 2001. - 256.

Hnatyshyn A.M., Shparyk Y.S., Position and tasks of remote sensing (RS) according to the requirements of Derzhheolkarty. - 200

Kolovsky M.Z., Evgrafov A.N., Semenov Yu.A., Slousch A.V., Advanced Theory of Mechanisms and Machines. - Springer – Verlag, 2000, 394 p.

Kuncevych V.M. Adaptive control to indeterminate dynamic objects // Adaptive control to dynamic objects. - Kiev: Science thought, 1982.

Miroshnik I.V., Fpadkov A.L., Nikiforov V.O., Nonlinear and adaptive control of complex dynamic systems. - St. Petersburg.: Nauka, 2000. - S.653.

Nair R., Maddocks J.H., On the forward kinematics of the parallel manipulators. - The International Journal of Robotics Research, Vol. 13, No. 2, April 1994, pp. 171 – 188.

Omata S., Khalid M., Rubiya Y., Neuro-control and its applications. - M: Radio, 2000. 272.

Reshetnev M.F. and others. Control and navigation satellites in circular orbits. – M.: Engineering, 1988

Sich-2 Space System: Tasks and Application Areas – K.: SSAU, 2011, - 48 p. – Ukr. and Eng.

Shikin E., Boreskov A. Computer Graphics. Dynamic, realistic images. - M.: "Dialog MIFI", 1995- 288p.

Stewart D., A Platform with Six Degrees of Freedom. - UK Institution of Mechanical Engineers Proceedings 1965-66, Vol 180, Pt 1, No 15.

Zaichenko Y.P., Fundamentals of intelligent systems. - K.: Publishing House "Word", 2004. - 352

6

Atmospheric Propagation of Terahertz Radiation

Jianquan Yao, Ran Wang, Haixia Cui and Jingli Wang
Tianjin University
China

1. Introduction

Terahertz (THz) radiation, sandwiched between traditional microwave and visible light, is the electromagnetic spectrum with the frequency defined from 0.1 to 10 THz ($1THz=10^{12}Hz$). Until recently, due to the difficulty of generating and detecting techniques in this region, THz frequency band remains unexplored compared to other range and tremendous effort has been made in order to fill in "THz gap" . (Zhang & Xu, 2009)

Recent advances provide new opportunities and widespread potential applications of THz in information and communication technology (ICT), material identification, imaging, non-destructive examination, global environmental monitoring as well as many other fields. The rapid development can be attributed to the nature of terahertz radiation, which offers the advantages of both microwave and light wave. The characteristics of THz atmospheric propagation now rank among the most critical issues in the principal application of space communication and atmospheric remote sensing. (Tonouchi, 2007)

Terahertz communication will benefit from the high-bit-rate wireless technology which takes advantage of higher frequency and broader information bandwidth allowed in this range than microwave. It is possible for such a system to achieve data rate in tens of gigabits per second. (Lee, 2009) However, as shown in Figure 1, the atmospheric opacity severely limits the communication applications at this range (Siegel, 2002) and it is the commercial viability rather than technological issues that will undoubtedly determine whether THz communication will be carried out into practical application.

The overview of the THz remote sensing from the National Institute of Information and Communications Technology (NICT) in Japan is given in Figure 2. (Yasuko, 2008) Many biological and chemical compounds exhibit distinct spectroscopic response in THz range, which presents tremendous potential in the environmental monitoring of atmospheric chemical compositions (water, oxygen, ozone, chlorine and nitrogen compounds, etc.) and the identification of climate evolution in the troposphere and lower stratosphere. (Tonouchi, 2007) The knowledge about atmospheric attenuation will illustrate the optimum frequency bands for sensing systems while the material database will discriminated atmospheric components.

Based on these considerations, there are three fundamental problems as follow: (Foltynowicz et al., 2005) (1)To confirm the atmospheric transparency in the THz range and

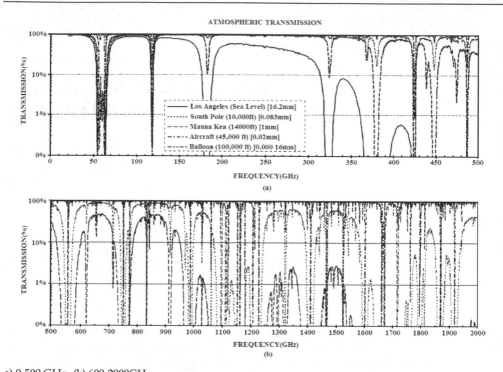

a) 0-500 GHz, (b) 600-2000GHz

Fig. 1. Atmospheric transmission in the terahertz region at various locations and altitudes

Fig. 2. Overview of NICT THz remote sensing

find out the air transmission windows for communicating and sensing system. (2)To collect the spectroscopic fingerprinting of atmospheric molecules for Terahertz atmospheric monitoring. (3)To improve the signal to noise ratio and restore the original signal from the observed signal by the process of deconvolution. (Ryu and Kong, 2010) It is essential to understand the actual effects on the amplitude and phase of THz radiation propagating through the atmosphere, which depends on the frequency of incident wave, gas components, and ambient temperature or barometric pressure in different atmospheric conditions.

This chapter aims to provide the theoretic instructions for the applications above and illuminate characteristics of THz atmospheric propagation. The fundamental theory has been systematically introduced, with the physical process of Lamber-beer law, Mie scattering theory and so on. The atmospheric absorption, scattering, emission, refraction and turbulence are taken into account and a special focus is put on the detailed derivation and physical significance of radiative transfer equation. Additionally, several THz atmospheric propagation model, including Moliere, SARTre and AMATERASU, are introduced and compared with each other. The conclusions are drawn by giving the future evolutions and suggestions of further study in this region.

2. Fundamental theories of terahertz atmospheric propagation

The framework of fundamental physical concepts and theories in the process of THz atmospheric propagation is shown in Figure 3. The three fundamental physical concepts (atmospheric extinction, atmospheric emission and background radiation) on the left can be uniformly expressed in the radiative transfer equation, which is the foundation of THz atmospheric propagation mode and describes the processes of energy transfer along a given optical path. Other elements (atmospheric refraction and turbulence) results in a correction and optimization of the integration path-length and radiative transfer algorithm in practical solution procedure.

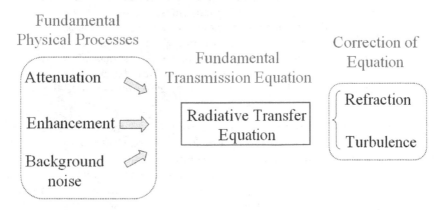

Fig. 3. The fundamental physical concepts and theories

2.1 Fundamental physical processes

2.1.1 Atmospheric extinction

In the process of the interaction between electromagnetic wave and medium, THz radiation is attenuated by absorption as well as scattering out of their straight path. The atmospheric extinction is illustrated by Lamber-beer law and mainly causing the energy attenuation of incident wave. The differential and integral forms of the mathematical expression is

$$dI(v) = -\alpha_v(z)I(v)dz \qquad I_{r_1}(v) = I_{r_0}(v)e^{-\int_{r_0}^{r_1} \alpha_v(z)dz} \tag{1}$$

$I_{r0}(v)$ denotes the incident radiance entering the optical path (r_0,r_1) at the frequency v and $I_{r1}(v)$ is the outgoing radiance. The opacity or optical thickness is defined as

$$\tau_v(r_0,r_1) = \int_{r_0}^{r_1} \alpha_v(z)dz \tag{2}$$

and the transmission is

$$\eta_{r_0,r_1} = \frac{I_{r_1}}{I_{r_0}} = e^{-\tau_v(r_0,r_1)} \tag{3}$$

Extinction coefficient $\alpha_v(z)$ can be expressed mathematically as the summation of the absorption and scattering coefficient, α_a and α_s, separately

$$\alpha_e = \alpha_a + \alpha_s \tag{4}$$

The atmospheric absorption, particularly from water vapor, involves the linear absorption and continuum absorption, while the atmospheric scattering mainly depends on aerosols.

2.1.1.1 The absorption of water vapor

The linear and continuum absorption constitutes the THz atmospheric absorption, which is dominated by water vapor. The former is comprised most of the absorption lines in the air, which is due to the molecular rotational transitions. The absorption lines of water vapor are characterized by spectroscopic parameters, including the center frequency, oscillator intensity, and pressure broadening coefficient. (Yasuko and Takamasa, 2008) Most of these optical properties have been conveniently catalogued into databases, such as JPL (Jet Propulsion Laboratory) and HITRAN (Rothman et al., 2009) to stimulate the line by line absorption.

The atmospheric absorption spectrum doesn't correspond to the accumulation of water vapor absorption lines. The continuum absorption is what remains after subtraction of linear contributions from the total absorption that can be measured directly. (Rosenkranz, 1998) It may be observed in wide electromagnetic spectrum (from microwave to infrared) and cannot be described by water vapor absorption lines. Its generating mechanism is not sufficiently understood while several theories have been proposed, including anomalous far-wing absorption, (Ma and Tipping, 1992) absorption by dimmers and larger clusters of water vapor, and absorption by collisions between atmospheric molecules. (Ma and Tipping, 1992) A semi-empirical CKD model is applicable in a wide frequency range and has been proven successful in some aspects. (Clough et al., 1989) For the simulation at frequencies below 400GHz, Liebe model could be used for dry air and water vapor continua. (Liebe, 1989) Figure 4 illustrates the discrepancy between radio-wave and infrared wave propagation models. The radio-wave model is calculated with JPL line catalog and

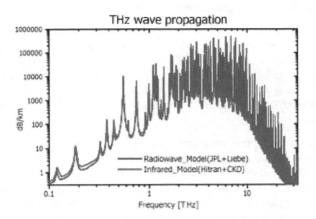

Fig. 4. The linear and continuum absorption of THz wave from NICT

Liebe model for continuum absorption while the infrared model is on the basis of HITRAN line catalog and CKD continuum model. (Yasuko and Takamasa, 2008)

2.1.1.2 The scattering of aerosol

In parallel, scattering effect also results in the energy attenuation along the optical path. It comprises the molecular Rayleigh scattering and the Mie scattering by aerosols and water vapor coagulum. As the wavelength of THz radiation lies in the order of aerosols, only Mie scattering should be taken into consideration. Aerosol particles mainly refer to the solid and liquid particles suspending in the atmosphere, for example, dusts, salts, ice particles and water droplets, and the Mie scattering effect mainly depends on their size-distribution, complex refractive index and the wavelength of incident radiation.

It is difficult to simulate the scattering by aerosols due to their large scale change in time and space domain. The scale distribution is an important concept to describe aerosols, and the spectrum pattern commonly includes:

2.1.1.2.1 Revision spectrum

$$\frac{dN(r)}{dr} = ar^{\alpha} \exp(-b^{\gamma}) \qquad (5)$$

Where N is granule number in the unit volume, r is the radius of particle, a, b, a, γ is the constant which depends the origin of aerosol, including Mainland (Haze L), Sea (Haze M) and High Stereotype (Haze H).

2.1.1.2.2 Junge spectrum

$$\frac{dN}{d\log r} = cr^{-v} \qquad (6)$$

In the expression above, v is the spectrum parameter, usually taking 2~4. The parameter c relates to the total density of aerosols.

2.1.1.3 Terahertz spectroscopic measurement technology

The THz spectroscopic parameters above will directly influence the accuracy of atmospheric propagation model and should be precisely measured in laboratory experiments. Currently, Terahertz Time-domain Spectroscopy (THz-TDS) technology and Fourier-transform Infrared Spectroscopy (FT-IR) have attracted a great deal of attention. A typical THz-TDS arrangement includes a femtosecond (fs) laser, a THz emitter source, a THz detector, focusing and collimating parts, a motorized delay line, a lock-in amplifier, and a data acquisition system.

As shown in Figure 5, the femtosecond laser is split into THz generation and detection arms. Coming from the same source, the pump and probe pulses have a defined temporal relationship. The THz radiation is excited by focusing the pulse onto a photoconductive antenna and the emitted THz pulses are collimated and focused onto the sample by a pair of parabolic mirrors; samples can be scanned across the focus to build up a two-dimensional image, with spectral information recorded at each pixel. (Baxter, 2011) The reflected or transmitted THz pulse is then collected and focused with another pair of parabolic mirrors onto a detector, which is a second photoconductive antenna or a sampling electro-optical crystal. The probe beam is measured with a quarter wave-plate, a Wollaston polarization (WP) splitting prism, and two balanced photodiodes. Lock-in techniques can be used to measure the photodiode signal with the modulated bias field of the photoconductive emitter as a reference. Furthermore, by measuring the signal as a function of the time delay between the arrival of THz and probe pulses, the THz time-domain electric field can be reconstructed. A computer controls the delay lines and records data from the lock-in amplifier, and the Fourier transform expresses the frequency spectrum of THz radiation. (Davies el al., 2008)

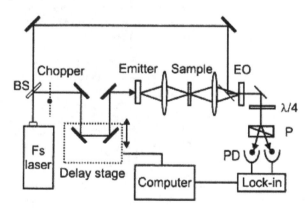

Fig. 5. Schematic experimental setups for THz-TDS system

Fourier transform infrared (FTIR) spectroscopy is a technique to obtain an infrared spectrum of absorption, emission, photoconductivity or Raman scattering of the samples. It consists of an incoherent high-pressure mercury arc lamp, a far-IR beam splitter (free-standing wire grid or Mylar), focusing and collimating optical parts for far infrared, a thermal detector, a motorized delay line, and a data acquisition system, just as Figure 6 plots. The source is generated by a broadband light source containing the full spectrum of wavelengths. The

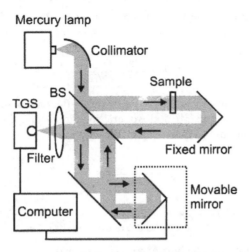

Fig. 6. Schematic experimental setups for far-IR Fourier transform spectroscopy

light shines into a Michelson interferometer, that allows some wavelengths to pass through but blocks others due to wave interference. Computer processing is required to turn the original data into the desired result.

Compared to other spectroscopic techniques, THz-TDS presents a series of advantages. THz pulse has ps pulse duration, resulting in the intrinsic high temporal resolution and is very suitable for the dynamic spectroscopic measurement. THz-TDS provides coherent spectroscopic detection and a direct record of the THz time-domain pulse. It enables the determination of the complex permittivity of a sample, consisting of the amplitude and phase, without the requirement of Kramers-Kronig relationship. (Zhang & Xu, 2009) Additionally, time-gating technology in sampling THz pulses has been employed, which dramatically suppresses the background noise. It is especially useful to measure spectroscopy with high background radiation which is comparable or even stronger than the signal. In terms of signal-to-noise ratio, THz-TDS is advantageous at low frequencies less than 3 THz, while Fourier transform spectroscopy works better at frequencies above 5 THz. (Han et al., 2001)

2.1.2 Atmospheric emission

THz radiation propagating in the atmosphere also experiences the process of enhancement. THz emission is defined as source term J, comprising the thermal emission J_B and the scattering source term J_S. Compared with the attenuation by scattering out of the line-of-sight, scattering into the path is considered as a source of radiation as well, including the source sole scattering on direct radiation condition J_{SS} and the multiple scattering source J_{MS}. (Mendrok, 2006) The expression of source terms is

$$J = J_B + J_S = J_B + J_{SS} + J_{MS} \qquad (7)$$

The thermal emission term is defined as

$$J_B = (1 - \omega_0)B(T) \qquad (8)$$

$B(T)$ denotes the Planck emission term which is given by Planck's function describing the radiation of a black-body at temperature T:

$$B_v(T) = \frac{2hv^3}{c^2} \frac{1}{e^{hv/k_B T} - 1}$$
(9)

where h is Planck's constant, c the speed of light, and k_B denotes Boltzmann's constant. w_0 is the scattering albedo of the "mixed" atmospheric medium along the line-of-sight, which is calculated from molecular and particle optical properties:

$$\omega_0 = \frac{\alpha_s^{par}}{\alpha_s^{par} + \alpha_a^{par} + \alpha_s^{mol}}$$
(10)

where α_s and α_a are scattering and absorption coefficients with superscripts 'mol' and 'par' denoting properties of molecular and particulate matter, respectively.

The scattering source term into the optical path is described as:

$$J_s(\Omega) = \frac{\alpha_s}{\alpha_e} \frac{1}{4\pi} \int_0^{4\pi} P(\Omega, \Omega') I(\Omega') d\Omega'$$
(11)

It comprises radiation incident from all directions Ω' scattered into the direction of interest Ω. While the scattering coefficient α_s accounts for the scattered fraction of radiation, the phase function $P(\Omega, \Omega')$ can be interpreted as the probability of incident radiation being scattered from direction Ω' into direction Ω with the normalizing condition:

$$\frac{1}{4\pi} \int_0^{4\pi} P(\Omega, \Omega') d\Omega' = 1$$
(12)

$I(\Omega')$ describes the incident radiation field in terms of incident direction for the calculation of the scattering source term.

2.1.3 Background radiation

Remote observations of the atmosphere can be performed at different geometries, as Figure 7 shows. The case that the line-of-sight goes through a long tangential atmospheric path above the ground is commonly referred to as limb-sounding geometry. If the line-of-sight crosses the surface, it is called nadir-sounding geometry. The up-looking case can be obtained by inverting the sense of the nadir observation. The background radiation of THz wave in the atmosphere mainly results from many kinds of electromagnetic radiation in the interstellar space or from the planet surface. For limb-sounding and up-looking, it is the cosmologic radiation at 3K, and for nadir-sounding (or down-looking), it is the earth surface emission.

2.2 Radiative transfer equation

Radiative transfer is the physical phenomenon of energy transferring in the form of electromagnetic radiation. The propagation of radiation through a medium is affected by the

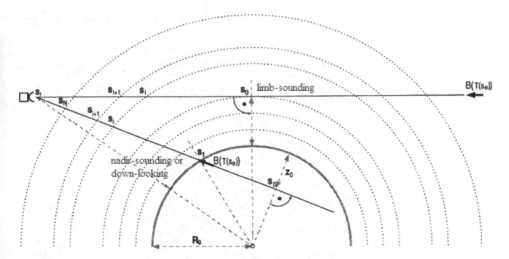

Fig. 7. Geometry inlcuding limb-sounding and nadir-sounding

three concepts (attenuation, enhancement, and background radiation) occurring along the line-of-sight and the equation of radiative transfer describes these interactions mathematically. It is the foundation of THz atmospheric propagation model, and the derivation is as follow: (Thomas & Stamnes, 2002)

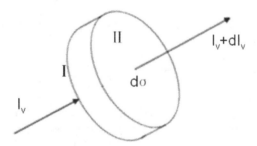

Fig. 8. The input and output optical intensity

The fundamental quantity which describes a field of radiation is the spectral intensity. Let's think of a very small area element in the radiation field, as the Figure 8 above, the radiant energy of incident light in the surface I of an infinitesimal volume is:

$$dE^{in} = I_v d\omega dv d\sigma dt \tag{13}$$

where I_v is radiant intensity, dw solid angle, dv frequency interval, $d\sigma$ basal area, and dt denotes the time of radiation (polarization will be ignored for the moment). And the emergent radiant energy from surface II is:

$$dE^{out} = (I_v + dI_v) d\omega dv d\sigma dt \tag{14}$$

According to the Lamber-beer law, with the absorption coefficient α_v, the radiant energy absorbed by the medium is:

$$dE_\alpha = -\alpha_v dE^{in} dr = -\alpha_v I_v d\omega dv d\sigma dt dr \tag{15}$$

With the emission coefficient j_v, the radiant energy of medium emission is:

$$dE_e = j_v d\omega dv d\sigma dt dr \tag{16}$$

In accordance with energy conservation law, we get:

$$dE^{out} = dE^{in} + dE_e + dE_\alpha \tag{17}$$

Substituting equation (8)~(11) into equation (12):

$$dI_v d\omega dv d\sigma dt = j_v d\omega dv d\sigma dt dr + (-\alpha_v I_v) d\omega dv d\sigma dt dr \tag{18}$$

A particularly useful simplification of the radiative transfer equation occurs under the conditions of local thermodynamic equilibrium (LTE). In this situation, the atmosphere consists of massive particles which are in equilibrium with each other, and therefore have a definable temperature. For the atmosphere in LTE, the emission coefficient and absorption coefficient are functions of temperature and density only, and the source function is defined as $S_v \equiv j_v / \alpha_v$. It equals the Planck function according to Kirchhoff's law:

$$S_v \equiv j_v / \alpha_v = B_v(T) \tag{19}$$

Given the definition of opacity or optical thickness: $d\tau_v = \alpha_v dr$, we get the differential form of radiative transfer equation from equation (18):

$$\frac{dI_v}{d\tau_v} = S_v - I_v \tag{20}$$

To solve this single-order partial differential equation along integral path (r_0, r_1), with the integral variable r, we get the integral form of radiative transfer equation:

$$I_v(r_1) = I_v(r_0) e^{-\int_{r_0}^{r_1} \alpha_v(r) dr} + \int_{r_0}^{r_1} e^{-\int_r^{r_1} \alpha_v(r') dr'} S_v(r) \alpha_v(r) dr \tag{21}$$

Under the assumption of LTE, the equation can be written as:

$$I_v(r_1) = I_v(r_0) e^{-\int_{r_0}^{r_1} \alpha_v(r) dr} + \int_{r_0}^{r_1} B_v(T) \alpha_v(r) e^{-\int_r^{r_1} \alpha_v(r') dr'} dr \tag{22}$$

The physical significance of radiative equation lies in the processes of absorption and emission of atmosphere at the position r along a given optical path (r_0, r_1), with the first term on the right side describing the background radiation attenuated by atmosphere while the second one standing for atmospheric emission and absorption. $I_v(r_1)$ is the outgoing radiance arriving the sensor at the frequency v and $I_v(r_0)$ corresponds to the background radiance entering the optical path.

As the radiative transfer equation results from energy conservation law, it is applicable to the whole electromagnetic spectrum, from radio wave to visible light. In the course of this work, radiation has only been discussed in terms of scalar intensity. Considering the polarization, the radiation is described by four components (I, Q, U, V) of the Stokes vector and a complete description of interaction between the medium and the radiation will be expressed. However, scalar radiative transfer is usually a good approximation for most situations in radiative transfer modeling.

2.3 Elements to promote the algorithm

2.3.1 Atmospheric turbulence

Turbulence is a flow regime characterized chaotically and stochastically, the problems of which are thus treated statistically rather than deterministically. The turbulent atmospheric optical property is changing with the temporal and spatial variation, resulting in the fluctuation of atmospheric refractive index. The essence of turbulence effect is the influence of medium disturbance on the transmission of incident THz radiation, including the beam drift, jitter, flickering, distortion, and degeneracy of the spatial coherence.

The turbulent consequence mainly depends on the relationship of turbulent scale l and the characteristic dimension of the incident radiation d_B.

On condition that $l>>d_B$, THz beam deflects during the process of the propagation in turbulence and mainly cause beam drifting on the receiver. When turbulent scale l is equal to the characteristic dimension d_B, the light beam will also experience stochastic deflection, resulting in the image spot jitter. If $l<<d_B$, the influence of scattering and diffraction leads to the intensity flickering of THz beam. (Yao & Yu 2006)

Additionally, in terms of incident radiation, fully coherent light beams are sensitive to the properties of the medium through which they are propagating and the turbulence-induced spatial broadening is the major limiting factor in most applications. Partially coherent beams are less affected by atmospheric turbulence than fully ones. (Shirai 2003)

2.3.2 Atmospheric refraction

The atmospheric refraction results from the uneven distribution of air in horizontal and vertical directions. When passing through the atmosphere, the line of sight is refracted and bended towards the surface of the planets. Taking refraction into account will correct and promote the radiative transfer path with some elementary geometrical relationships, as plotted in Figure 9.

In conclusion of Section 2, the general idea to solve these problems above is to study the various effects independently and superpose them. Currently, most researches are mainly focused on the atmospheric extinction and the establishment of radiative transfer model.

3. THz atmospheric propagation model

3.1 Moliere

Microwave Observation Line Estimation and Retrieval (Moliere), developed at the Bordeaux Astronomical Observatory (France), is the versatile forward and inversion model for

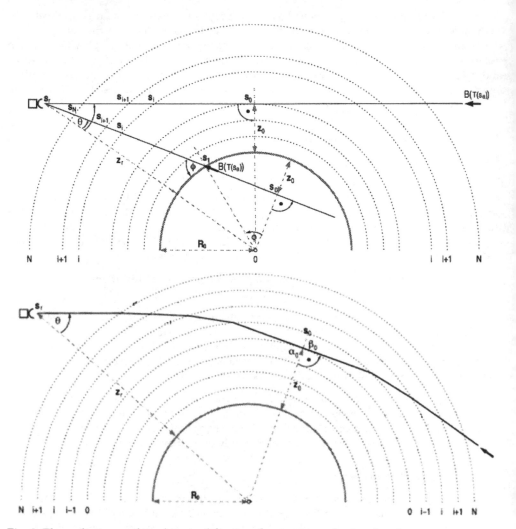

Fig. 9. The radiation path and its modification due to atmospheric refraction

millimeter and sub-millimeter wavelength observations on board the Odin satellite, including a non-scattering radiative transfer model, a receiver simulator and an inversion code. The forward models comprise spectroscopic parameters, atmospheric radiative transfer model, and instrument characteristics in order to model and compute the searched atmospheric quantities. In parallel, inversion techniques have been developed to retrieve geophysical parameters such as temperature and trace gas mixing ratios from the remotely measured spectra. (Urban et al., 2004)

Moliere is presently applied to data analysis for ground-based and space-borne heterodyne instruments and definition studies for future limb sensors dedicated to Earth observation and Mars exploration. However, this code can not be used when both up-looking and down-looking geometries should be considered together, and for limb geometry if the receiver is inside the atmosphere, such as balloon and airplane.

3.2 SARTre

The new radiative transfer model [Approximate] Spherical Atmospheric Radiative Transfer model (SARTre) has been developed to provide a consistent model that accounts for the influence of aerosols and clouds, e.g. water droplets or ice particles. It includes emission and absorption as well as scattering as sources/sinks of radiation from both solar and terrestrial sources in the spherical shell atmosphere and is able to analyze data measured over the spectral range from ultraviolet to microwaves. (Mendrok et al., 2008) SARTre is designed for monochromatic, high spectral resolution forward modeling of arbitrary observing geometries, especially for the limb observation technique.

The line-by-line calculation of molecular absorption cross sections has been adapted from the radiative transfer package MIRART (Modular Infrared Atmospheric Radiative Transfer). And the DISORT (Discrete Ordinate Radiative Transfer Model) package is used for the calculation of the incident radiation field when taking multiple scattering into account, under the assumption of a locally plane-parallel atmosphere. (Mendrok et al., 2008)

3.3 AMATERASU

The Advanced Model for Atmospheric Terahertz Radiation Analysis and Simulation (AMATERASU) is developed by the National Institute of Information and Communications Technology (NICT) THz project. This project aims to develop THz technology for various applications concerning the telecommunications, atmospheric remote sensing to retrieve geophysical parameters and the study of the thermal atmospheric emission in the Earth energy budget. The framework of AMATERASU has been shown in Figure 10, mainly consisting of the spectroscopic parameters and the radiative transfer equation, as mentioned above.

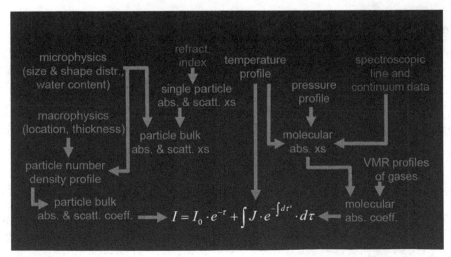

Fig. 10. The framework of AMATERASU from NICT

The AMATERASU has a strong heritage from the two models above, respectively in the non-scattering and scattering case. The first stage concerns a non-scattering and homogeneous atmosphere, based on the original Moliere receiver simulator and retrieval codes. The absorption coefficient module has been extent to THz region and a more general radiative transfer module has been implemented to handle different geometries of optical paths and any location for the receiver. (Baron et al., 2008) The advanced version has taken the scattering effect into consideration. Modules related to optical properties of atmospheric particles and to scattering have been adapted from SARTre. The complex refractive index data of aerosols in THz region should be emphasized as a crucial parameter for radiative transfer algorithms. (Mendrok et al., 2008)

As for the practical applications, the THz atmospheric propagation models above should be compared with each other and validated against the real laboratory measurements in order to verify the data accuracy and correctness of the algorithm hypothesis. (Wang et al., 2011)

4. Conclusion

In this chapter, we have discussed the fundamental theory in the process of THz atmospheric propagation. Several kinds of THz atmospheric propagation models have been introduced as well. The critical issues lie in the construction of radiative transfer algorithm, the collection of accurate spectral parameters, such as linear and continuum absorption and complex refractive index in THz region, and the standardization of measurement procedures. The ultimate objective is to construct the atmospheric propagation model in different kinds of climatic conditions on the basis of the theoretical analysis and the material database.

5. Acknowledgment

This program is supported by the National Basic Research Program of China under Grant No. 2007CB310403.

6. References

Baron P.; Mendrok J. & Yasuko K. (2008). AMATERASU: Model for Atmospheric TeraHertz Radiation Analysis and Simulation. Journal of the National Institute of Information and Communications Technology, Vol. 55, No. 1, (March 2008), pp. 109-121

Baxter J. & Guglietta G. (2011). Terahertz Spectroscopy. Analytical Chemistry, Vol. 83, No. 12, (June 2011), pp. 4342-4368

Clough S.; Kneizys F. & Davies R. (1989). Line shape and the water vapor continuum. Atmospheric Research, Vol.23, No.3, (October 1989), pp. 229-241

Davies A.; Burnett A. & Fan W. (2008). THz spectroscopy of explosives and drugs. Materialstoday, Vol. 11, No. 3, (March 2008), pp. 18-26, ISSN 1369-7021

Foltynowicz, R.; Wanke, M. & Mangan, M. (2005). Atmospheric Propagation of THz Radiation, Sandia National Laboratories, New Mexico, America

Han P.; Tani M. & Usami M. (2001). A direct comparison between terahertz time-domain spectroscopy and far-infrared Fourier transform spectroscopy. Journal of Applied Physics, Vol. 89, No. 4, (February 2001), pp. 2357-2359, ISSN 0021-8979

Lee, Y. (2008). Principles of Terahertz Science and Technology, Springer Science+Business Media, ISBN 978-0-387-09539-4, New York, America.

Liebe H. (1989). MPM-An atmospheric millimeter-wave propagation model. International Journal of Infrared and Millimeter Waves, Vol.10, No.6, (February 1989), pp. 631-650

Ma Q. & Tipping R. (1992). A far wing line shape theory and its application to the foreign-broadened water continuum absorption. Journal of Chemical Physics, Vol.97, No.2, (April 2008), pp. 818-828, ISSN 0021-9606

Ma Q. & Tipping R. (1999). The averaged density matrix in the coordinate represesntation: application to the calculation of the far-wing line shapes for H2O. Journal of Chemical Physics, Vol.111, No.13, (June 1999), pp. 5909-5921, ISSN 0021-9606

Mendrok J. (2006). The SARTre Model for Radiative Transfer in Spherical Atmospheres and its application to the Derivation of Cirrus Cloud Properties, Freie Universität, Berlin, Germany

Mendrok J.; Baron P. & Yasuko K. (2008). The AMATERASU Scattering Module. Journal of the National Institute of Information and Communications Technology, Vol. 55, No. 1, (March 2008), pp. 123-132

Rosenkranz P. (1998). Water vapor microwave continuum absorption: a comparison of measurements and models. Radio Science, Vol.33, No.4, (July 1998), pp. 919-928

Rothman L.; Gordon I. & Barbe A. (2009). The HITRAN 2008 molecular spectroscopic database. Journal of Quantitative Spectroscopy and Radiative Transfer, Vol.110, No.9, (June 2009), pp. 533-572

Ryu, C. & Kong, S. (2010). Atmospheric degradation correction of terahertz beams using multiscale signal restoration. Applied Optics, Vol.49, No.5, (February 2010), pp. 927-935

Shirai T. (2003). Mode analysis of spreading of partially coherent beams propagating through atmospheric turbulence. *Journal of the Optical Society of America A*, pp. 1094-1102

Siegel, P. (2002). Terahertz Technology. IEEE Transactions on microwave theory and techniques, Vol.50, No.3, (March 2002), pp. 910-928, ISSN 0018-9480

Thomas G. & Stamnes K. (2002). Radiative Transfer in the Atmosphere and Ocean, Press Syndicate of the University of Cambridge, ISBN 0-521-40124-0, Cambridge, United Kingdom

Tonouchi, M. (2007). Cutting-edge terahertz technology. Nature Photonics, Vol.1, No.2, (February 2007), pp. 97-105, ISSN 1749-4885

Urban J.; Baron P. & Lautié N. (2004). Moliere(v5): a versatile forward-and inversion model for the millimeter and sub-millimeter wavelength range. Journal of Quantitative Spectroscopy & Radiative Transfer, Vol. 83, No. 4, (February 2004), pp. 529-554, ISSN 0022-4073

Urban J.; Baron P. & Lautié N. (2004). Moliere(v5): a versatile forward-and inversion model for the millimeter and sub-millimeter wavelength range. Journal of Quantitative Spectroscopy & Radiative Transfer, Vol. 83, No. 4, (February 2004), pp. 529-554, ISSN 0022-4073

Wang R.; Yao J. & Xu D. (2011). The physical theory and propagation model of THz atmospheric propagation. Journal of Physics: Conference Series, Vol. 276, No. 1. (March 2011), pp. 012223, ISSN 1742-6596

Yao, J. & Yu Y. (2006). Optoelectronic Technology, Higher Education Press, ISBN 7-04-019255-1, Bei Jing, China

Yasuko, K. (2008). Terahertz-Wave Remote Sensing. Journal of the National Institute of Information and Communication Technology, Vol.55, No.1, (March 2008), pp. 79-81

Yasuko K. & Takamasa S. (2008). Atmospheric Propagation Model of Terhertz-Wave. Journal of the National Institute of Information and Communications Technology, Vol.55, No.1, (March 2008), pp. 73-77

Zhang, X. & Xu J. (2009). Introduction to THz Wave Photonics, Springer Science+Business Media, ISBN 978-1-4419-0977-0, New York, America

7

Hardware Implementation of a Real-Time Image Data Compression for Satellite Remote Sensing

Albert Lin

National Space Organization
Taiwan, R.O.C.

1. Introduction

The image data compression is very important to reduce the image data volume and data rate for the satellite remote sensing. The chapter describes how the image data compression hardware is implemented and uses the FORMOSAT-5 Remote Sensing Instrument (RSI) as an example. The FORMOSAT-5 is an optical remote sensing satellite with 2 meters Panchromatic (PAN) image resolution and 4 meters Multi-Spectrum (MS) image resolution, which is under development by the National Space Organization (NSPO) in Taiwan. The payload consists of one PAN band with 12,000 pixels and four MS bands with 6,000 pixels in the remote sensing instrument. The image data compression method complies with the Consultative Committee for Space Data Systems (CCSDS) standard CCSDS 122.0-B-1 (2005). The compression ratio is 1.5 for lossless compression, 3.75 or 7.5 for lossy compression. The Xilinx Virtex-5QV FPGA, XQR5VFX130 is used to achieve near real time compression. Parallel and concurrent handling strategies are used to achieve high-performance computing in the process.

2. Image compression methodology

The CCSDS Recommended Standard for Image Data Compression is intended to be suitable for spacecraft usage. The algorithm complexity is sufficiently low for hardware implement and memory buffer requirement. It can support strip-based input format for push broom imaging. The compressor consists of two functional blocks, Discrete Wavelet Transfer (DWT) and Bit Plane Encoder (BPE). The image compression methodology is described in the following sections.

2.1 Discrete wavelet transform

The CCSDS Recommendation supports two choices of DWT: an integer DWT (IDWT) and a floating point DWT (FDWT). The integer DWT requires only integer arithmetic, is capable of providing lossless compression, and has lower implementation complexity, but lower compression ratio. The floating point DWT provides improved compression effectiveness, but requires floating point calculations and cannot provide lossless compression.

The DWT stage performs three levels of two-dimensional (2-d) wavelet decomposition and generates 10 subbands as illustrated in Fig. 1. The low pass IDWT is as Equation (1) and the high pass IDWT is as Equation (2). The low pass FDWT is as Equation (3) and the high pass FWDT is as Equation (4), j=0, 1,...11999 for PAN band, j=0,1,...5999 for MS bands in FORMOSAT-5 case.

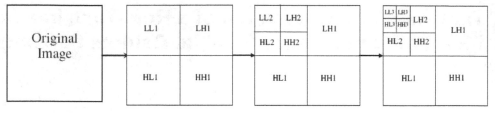

Fig. 1. Three-Level 2-d DWT Decomposition of an Image

$$C_j = \frac{1}{64}x_{2j-4} - \frac{1}{8}x_{2j-2} + \frac{1}{4}x_{2j-1} + \frac{23}{32}x_{2j} + \frac{1}{4}x_{2j+1} - \frac{1}{8}x_{2j+2} + \frac{1}{64}x_{2j+4} \tag{1}$$

$$D_j = \frac{1}{16}x_{2j-2} - \frac{9}{16}x_{2j} + x_{2j+1} - \frac{9}{16}x_{2j+2} + \frac{1}{16}x_{2j+4} \tag{2}$$

$$C_j = \sum_{n=-4}^{4} h_n X_{2j+1+n}; \qquad j = 0,1,...,11999 \tag{3}$$

$$D_j = \sum_{n=-3}^{3} g_n X_{2j+1+n}; \qquad j = 0,1,...,11999 \tag{4}$$

For FDWT, the coefficients in the equation (3) and (4) are listed in Table 1. The coefficients used in the FORMOSAT-5 are a little different from those defined in the CCSDS 122.0-B-1. Just 24 bits, not 32 bits, are used for these coefficients in the FORMOSAT-5 to save FPGA multiplexer resource.

i	FDWT Coefficients defined in CCSDS		FDWT Coefficients used in FORMOSAT-5	
	Low Pass Filter, h_i	High Pass Filter, g_i	Low Pass Filter, h_i	High Pass Filter, g_i
0	0.852698679009	- 0.788485616406	0.852698564529	- 0.788485646247
±1	0.377402855613	0.418092273222	0.377402901649	0.418092250823
±2	- 0.110624404418	0.040689417609	- 0.110624432563	0.040689468383
±3	- 0.023849465020	- 0.064538882629	- 0.023849487304	- 0.064538883647
±4	0.037828455507		0.037828445434	

Table 1. Coefficients of floating point DWT

2.2 Bit plane encoder

After DWT processing, the Bit Plane Encoder handles DWT coefficient for data compression. The Bit Plane Encoder encodes a segment of images from most significant bit (MSB) to least significant bit (LST). The BPE encoding uses less bits to express image data to achieve compression ratio. In CCSDS 122.0-B-1, the maximum number of bytes in the compressed segment can be defined to limit the data volume. The quality limit can be defined to constraint the amount of DWT coefficient information to be encoded.

The BPE performs DC and AC data encryption as the flow shown in Fig.2. In DC part data encryption, AC part maximum value of each block will be computed. Then, a scheme should be used to determine how many bits for "DC_MAX_Depth" and "AC_MAX_Depth" in this segment. In addition, the DC and AC optimized encryption type and value of W/8 blocks should be determined. Finally, the DC part data and W/8 AC_MAX data will be encrypted and the bit stream is transmitted to next stage. W is the pixel size per image line, e.g. W is 12,000 for PAN image and W is 6,000 for MS image in FORMOSAT-5.

In AC part data encryption, it consists of 5 stages. Data encryption and bit-out proceed block by block in each stage. The entropy coding scheme is used by data encryption. The stage 0 is for processing DC 3rd part data. The stage 1 is for processing Parent part coefficients in each block. The stage 2 is for processing Children part coefficients in each block. The stage 3 is for processing Grand-Children part coefficients in each block. The stage 4 is just concatenated stage 1, stage2 and stage 3 left data. After adding segment header, the compressed image data are finished.

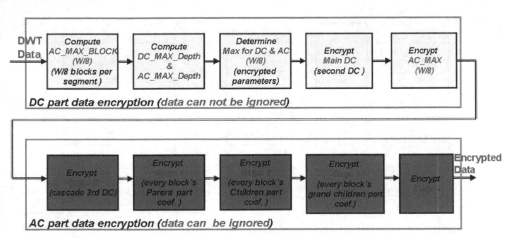

Fig. 2. BPE Encoding Flow

3. Hardware implementation

3.1 Architecture description

The image flow of the Remote Sensing Instrument in the FORMOSAT-5 is shown in Fig. 3. Behind the telescope, there is one CMOS sensor module inside the Focal Panel Assembly (FPA) to take the images. The CMOS sensor module can be accessed by two FPA electronics. The output data stream is sent to the Image Data Pre-processing (IDP) module in the RSI EU for data re-ordering. Then the resultant data are sent to the Image Data Compression (IDC) module for data compression. The compressed data with format header are stored in the Mass Memory (MM) modules under the control of the Memory Controller (MC) module. While the satellite flies above the ground station, the image files can be retrieved and transmitted to the ground station.

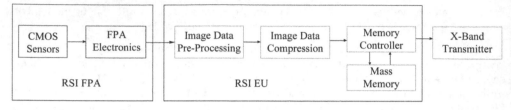

Fig. 3. Image Flow of Remote Sensing Instrument

3.2 Design and implementation

The image data input interfaces between each functional module are shown in the Fig. 4. The serial image data from FPA are re-ordered in the IDP to make the image data output in correct pixel order. Then the image data are transferred to IDC in parallel on 12-bit data bus with lower transmission clock rate. One channel of PAN data and four channels of MS data are compressed individually in the IDC. The compressed PAN and MS data are stored individually in image files under the control of MC module.

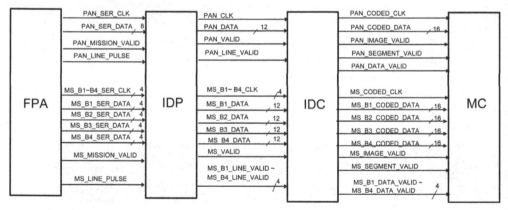

Fig. 4. Image Data Signal Interfaces between Functional Modules

3.3 Hardware design

The image data rate between each stage is shown in Fig. 5. The PAN sensors output are divided into 8 channels with 80Mbps rate individually to accommodate the high data rate. The channel rate for each MS band is 40Mbps. The parallel handling architecture can increase the image data handling speed.

The PAN and MS image data compression boards are shown in Fig. 6 a) & b). The architecture block diagram of the PAN channel in IDC is illustrated in Fig. 7. The MS channels are similar. The space grade Xilinx FPGA, XQR5VFX130, is used for image compression processing. The major characteristics of the XQR5VFX130 are 130,000 logic cells, 298 blocks of 36K bits RAM, 320 enhanced DSP slices,700Krad total dose, and etc. The PROM part for FPGA programming is XQR17V16, which has 16Mbits memory size with 50krad total dose capability. One XQR5VFX130 FPGA is used for PAN data compression.

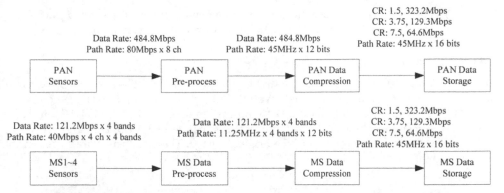

Fig. 5. Image Data Rate between each stage

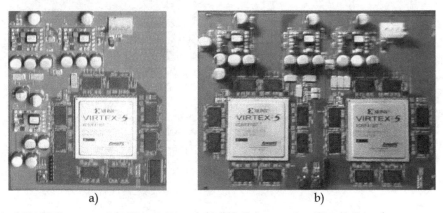

Fig. 6. a) PAN Compression Circuit Board; b) MS Compression Circuit Board

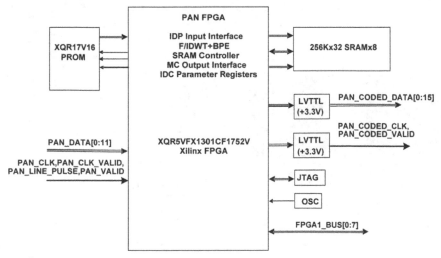

Fig. 7. Architecture Block Diagram of the PAN Channel in IDC

Two XQR5VFX130 FPGAs are used for four MS data compression. The external memories, 24 chips of 256K x 32 SRAM, are used as data buffer during compression process.

3.4 DWT process

The DWT flows at three levels are illustrated in Fig. 8a, 8b and 8c. The RAM memory banks are used for buffer storage. In the first level, the LL1, LH1, HL1 and HH1 are generated. Then, the LL1 is transmitted to level 2 DWT process to generate LL2, LH2, HL2 and HH2. The LL2 is transmitted to level 3 DWT process to generate LL3, LH3, HL3 and HH3. The LL3 contains the most information of the original image. These subbands are stored in the temporary buffers for BPE process.

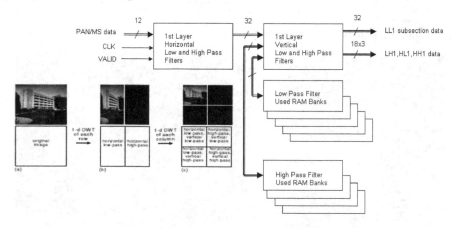

Fig. 8a. DWT Flow (1)

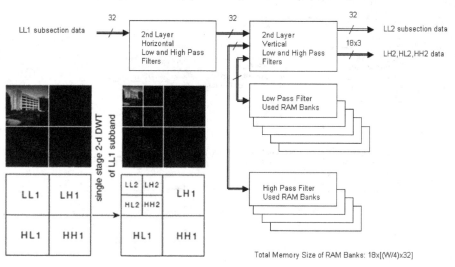

Fig. 8b. DWT Flow (2)

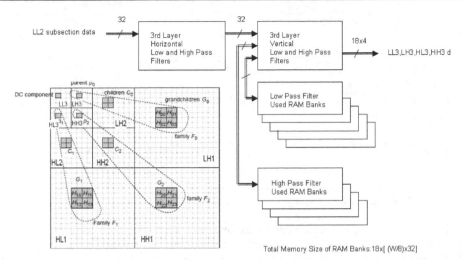

Fig. 8c. DWT Flow (3)

3.5 BPE process

The BPE module is the actual unit to perform data compression. When DWT acknowledges that one section data is completed and saved in the buffer, BPE retrieves the wavelet domain data from buffer and uses different compression scheme for different DWT sub-section data. According to various compression ratio requirements, BPE performs data truncation or appends zero fill bits. After necessary header information is added, the compressed data is sent to mass memory word by word for storage.

The compression data format is listed in Table 2. Within a segment, BitDepthDC is defined as the bit number of the maximum value in all DC coefficients. BitDepthAC is defined as the bit number of the maximum value in all AC coefficients. The amount of quantization q' of DC coefficients is determined by the dynamic range of the AC and DC coefficients in a segment in Table 3. DC quantization factor q is defined as q= max(q', BitShift(LL3)). The value of q indicates the number of least significant bits in each DC coefficient that are not encoded in the quantized DC coefficient values. The number of bits needed to represent each quantized DC efficient, N = max {BitDepthDC – q, 1}. For example, one segment has BitDepthDC=16 and BitDepthAC=4. According to Table 3, the DC quantization amount

Segment Header
Initial coding of DC coefficients
Coded AC coefficient bit depths
Coded bit plane b=BitDepthAC-1
Coded bit plane b=BitDepthAC-2
....
Coded bit plane b=0

Table 2. Compression Data Format

DC and AC dynamic range	q' value	Remark
$BitDepthDC \leq 3$	q' = 0	DC dynamic range is very small; no quantization is performed
$BitDepthDC - (1 + \lfloor BitDepthAC / 2 \rfloor) \leq 1$ and $BitDepthDC > 3$	q' = BitDepthDC-3	DC dynamic range is close to half the AC dynamic range
$BitDepthDC - (1 + \lfloor BitDepthAC / 2 \rfloor) > 10$ and $BitDepthDC > 3$	q' = BitDepthDC-10	DC dynamic range is much higher than half the AC dynamic range
Otherwise	$q' = 1 + \lfloor BitDepthAC / 2 \rfloor$	DC dynamic range is moderately higher than half the AC dynamic range

Table 3. DC Coefficient Quantization

q' = 16 -10 = 6. Then, DC quantization factor q is 6 and N = 16 - 6 =10. So, each DC coefficient bit(15) ~ bit(6) are encoded using coding quantization method, and bit(5) ~ bit(4) will just concatenated immediately at the end of the coded quantized DC coefficients of the segment, finally bit(3) ~ bit(0) are encoded at AC stage0 phase. The detailed coding algorithm is described in CCSDS 122.0-B-1 (2005).

The AC part data have the major portion of image (63/64), so AC part data coding dominates the whole compression performance. The CCSDS adopts bit plane encoding concept, that is, the most important bits of each AC subsection part data is encoded first, then less important bits, until specified segment byte limit size is achieved or bit 0 of each data segment is encoded. Even, it is needed to append zero bits to achieve segment byte limited size.

In order to have good compression efficiency, the CCSDS standard specifies AC Parent, Children, and Grand Children data to proceed entropy symbol mapping scheme. The basic concept of entropy coding is to use smaller bit pattern to represent more frequently repeated bit pattern.

In the CCSDS standard, a "gaggle" consists of a set of 16 consecutive blocks within a segment. There are two running phases in our design to use entropy coding scheme to represent the final coding result, pre-running phase and normal running phase. The pre-running phase is designed to get 2-bits、3-bits、and 4-bits entropy value for each gaggle on each bit-plane. The normal running phase is to use entropy table to map the final coding bits string. The detailed coding algorithm is described in CCSDS 122.0-B-1 (2005). The IDC implementation block diagram is shown in Fig. 9.

3.6 FPGA design optimization

Some design skills are used to save the limited multiplier and memory resources in the FPGA chip. In the Equation (1) and (2), nine multipliers for Low Pass Filter and seven multipliers for High Pass Filter are needed. Totally 3 x 2 x (9+7) = 96 multipliers are needed for 3 layers, horizontal and vertical, low pass and high pass filter. By using the multiplexers, adders and timing sharing algorithm in our IDC design as in Fig. 10 and 11, three multipliers for Low Pass Filter and two multipliers for High Pass Filter are needed. In other words, totally 3 x 2 x (3+2) = 30 multipliers are needed for 3 layers 2 dimension FDWT architecture, i.e. 66 multipliers are reduced.

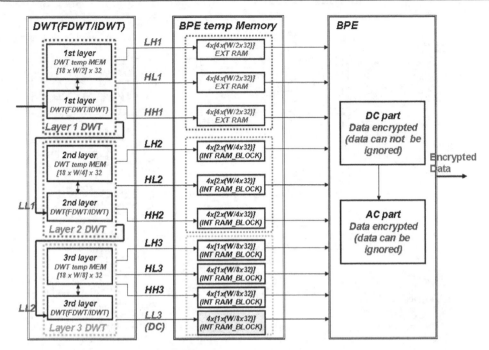

Fig. 9. IDC Implementation Block Diagram

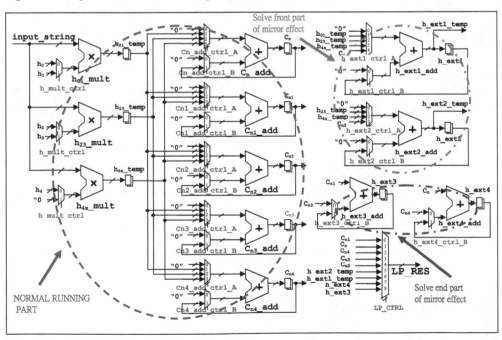

Fig. 10. Approach of 9 Taps Low Pass Filter in IDC

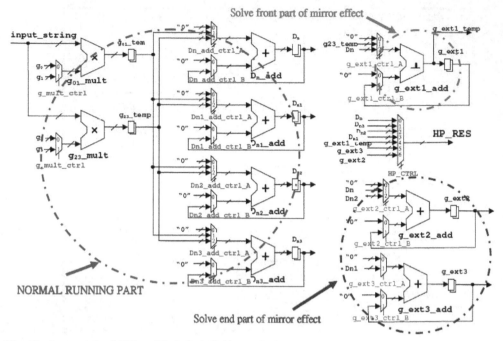

Fig. 11. Approach of 7 Taps High Pass Filter in IDC

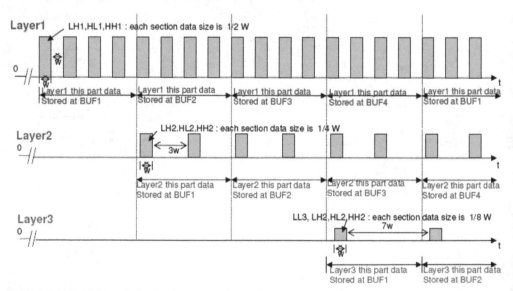

Fig. 12. DWT Timing Relation between Three Layers

The timing relation chart of DWT three layers is shown in Fig. 12. The "W" is the original source image width (pixels/line) which is 12000 for PAN and 6000 for MS in FROMOSAT-5.

The source clock is 45 MHz for PAN and 11.25 MHz for MS. In the Layer1, LH1, HL1 and HH1 data are generated every two source clocks with data size W/2 words. In the Layer2, LH2, HL2 and HH2 data are generated every four source clocks with data size W/4 words. In the Layer 3, LL3, LH3, HL3, HH3 data are generated every 8 source clocks with data size W/8 words. The data in different layers are generated interleavely to achieve high throughput for real time data processing.

The buffer size to handle the image compression is Width * Length for frame-based method. But for the strip-based method, just fixed buffer size, Width * 138, is needed. For 8 minutes FORMOSAT-5 PAN imaging data, the buffer size for frame-based will be 200,000 times of buffer size for strip-based. So, it is very important to use strip-based method to save memory size, cost, and handling time in satellite application, even for ground image handling. The total required memory can be reduced as shown in Table 4. It can save the cost and reduce the power consumption used by memory chips.

	CCSDS 120.1-G-1	FORMOSAT-5 Approach
Low Pass Filter	[2 x (9 x W/2n)] x 32 bits	[2 x (5 x W/2n)] x 32 bits
High Pass Filter	[2 x (7 x W/2n)] x 32 bits	[2 x (4 x W/2n)] x 32 bits

Where : W is pixels per line (12000); n is layer number (1~3)

Table 4. Memory Size in FDWT Implementation

The Xilinx Virtex-5QV FPGA static power is 2.49761 watts estimated by Xilinx XPower Analyzer tool. Since the throughput is 40.4 Msamples/sec for PAN, the power consumption of the compression FPGA is about 0.06 Watt/Msamples/sec. The total power consumption of the PAN compression board is about 5 watts, including SRAM and IO circuit, i.e. equivalent to 0.124 Watt/Msamples/sec.

There are some benefits to use space grade FPGA chip than ASIC. The space grade FPGA has good anti-radiation capability. The line pixel number and clock rate can be reconfigured. There are some comparisons of data compression chips in Table 5.

Features \ Model	FORMOSAT-5 RSI EU IDC	CAMBR DWT+BPE IC [Winterrowd 2009]	ANALOG DEVICES ADV202
Chip Type	Xilinx Space Grade FPGA	ASIC	ASIC
Compression Algorithm	CCSDS 122.0	CCSDS 122.0	JPEG2000
Line Width (Pixels)	12000	8192	4096
Bits Per Pixel	12	16	8, 10, 12, 14, 16
Input Data Rate	480Mbps	320Mbps	780Mbps
Radiation(Total Dose, Si)	700K	>=50K	Commercial
Power Consumption (Watt/Msamples/sec)	0.06	0.17	0.05

Table 5. Data Compression Chip Comparison

4. Image quality verification

The 12-bit test images in the CCSDS official website have been tested and similar results are gotten as in the CCSDS report. In order to consider more practical case, one North Vancouver image taken by FORMOSAT-2 satellite on 2009/12/9 is adopted. The compression ratios are set 1.5, 3.75 and 7.5. The Peak Signal to Noise Ratio (PSNR) is used as the performance index.

$$PSNR \equiv 20\log_{10}\frac{2^{B}-1}{\sqrt{MSE}}(dB),$$ (5)

where B denotes the bit depth and the Mean Squared Error (MSE) is given by

$$MSE = \frac{1}{w \cdot h}\sum_{i=1}^{w}\sum_{j=1}^{h}\left(x_{i,j}-\hat{x}_{i,j}\right)^{2}$$ (6)

where $x_{i,j}$ is the pixel of the original image, $\hat{x}_{i,j}$ is the pixel of the decoded image, w is the width of image and h is the height of image.

In our verification, one 8-lines strip-based segment is adopted with 1500 blocks for PAN and 375 blocks for MS. The average PSNR is calculated by Matlab® software. The test results are listed in the Table 6.

Compression Ratio	Image Methods	Panchro-matic Band	Red Band	Green Band	Blue Band	Infrared Band
CR=1.5	IDWT	Lossless	Lossless	Lossless	Lossless	73.1
	FDWT	51.1	51.1	51.1	51.1	51
CR=3.75	IDWT	47.3	47.8	44.3	45.3	41.1
	FDWT	47.7	48	45	45.8	41.7
CR=7.5	IDWT	43.1	41.8	37.6	38.1	38.5
	FDWT	43.6	42.2	38	38.5	35.1

* IDWT: Integer Discrete Wavelet Transform FDWT: Floating Point Discrete Wavelet Transform

Table 6. Image PSNR under Various Compressions

When the IDWT is used with compression ratio 1.5, the PSNR is very large to indicate near lossless compression, except the infrared band. When the FDWT is used with compression ratio 7.5, the PSNR may drop to 35dB which is worse than average PSNR 56.77dB using six 12-bit CCSDS test images. This is mainly because North Vancouver image shown in Fig. 13 is much more complicated than the standard CCSDS test images.

To use the satellite image as data input to real compression hardware, a set of simulated Focal Plane Assembly (FPA) is under development as illustrated in Fig. 14. The satellite image taken by FORMOSAT-2 is expanded from 8 bits to 12 bits per image pixel by adding random value of 4 least significant bits to simulate FORMOSAT-5 image. The test image can be downloaded from the personal computer to the image sensors simulator which is to replace the real image sensor array in the FPA. Then the test image can be transmitted out by the FPA simulator like real push broom image data. The test image will be compressed by hardware, then decompressed by software to check the hardware compression performance to simulated satellite image.

To have a quick check on hardware function, a test image with 1024 pixels x 1024 pixels size and 12 bits resolution has been downloaded to a prototype board. The test image is compressed by hardware, and then decompressed by software. These two images are shown

Fig. 13. North Vancouver Image Taken by FORMOSAT-2 satellite

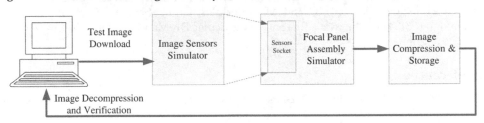

Fig. 14. Architecture of Image Compression Verification on Hardware

in Fig. 15. The PSNR is 82.8dB for compression ratio 1.5, 56.9dB for compression ratio 3.75, and 49.2dB for compression ratio 7.5.

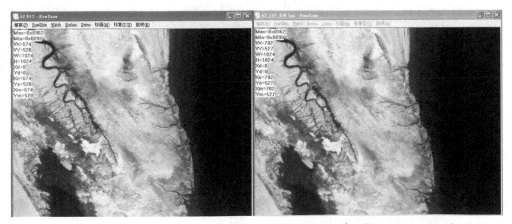

Fig. 15. Test Image before Compression (left) and Test Image after Decompression (right)

5. Conclusion

In this chapter there has been described the implementation of CCSDS recommended image data compression. The parallel processing, time sharing and computation via pure hardware in FPGA chip can achieve high-performance computing. The image data compression module based on FPGA has been developing to provide enough compression ratios with required image quality for FORMOSAT-5 mission. The performance has been verified by standard CCSDS 122.0 test images and FORMOSAT-2 images. The technology can be used on similar image data compression application in space. The compression throughput can be promoted following the improvement on the FPGA technology. The main advantage of this technique is that it allows real time image compression by efficient hardware implementation with low power consumption. This makes it especially suitable for satellite remote sensing.

6. Acknowledgment

The work is supported by the National Space Organization (NSPO) in Taiwan under the FORMOSAT-5 project. The author greatly acknowledge the following partners for their contribution: Dr. C. F. Change and Miss Cynthia Liu in NSPO on Image Algorithm Development and Image Quality Verification, CMOS Sensor Inc. on IDP module, Camels Vision Technologies Inc. on MC and MM modules, Chung-Shan Institute of Science & Technology (CSIST) on the whole EU, and in particular Dr. Mao-Chin Lin, Mr. Li-Rong Ran on IDC module.

7. References

CCSDS, "CCSDS 122.0 released 12-bits images", http://cwe.ccsds.org/sls/docs/sls-dc, (2007)

CCSDS, "Image Data Compression. Recommendation for Space Data System Standards", CCSDS 122.0-B-1. Blue Book. Issue 1. Washington, D.C., USA: CCSDS, (November 2005)

CCSDS, "Image Data Compression. Report Concerning Space Data Systems Standards", CCSDS 120.1-G-1. Green Book. Issue 1. Washington, D.C., USA: CCSDS, (June 2007)

Wang, Hongqiang, "CCSDS Image Data Compression C source codes", http://hyperspectral.unl.edu/, University of Nebraska-Lincoln, (Sept 2008)

Winterrowd, Paul, etc. "A 320 Mbps Flexible Image Data Compressor for Space Applications", IEEEAC paper#1311, 2009

8

Road Feature Extraction from High Resolution Aerial Images Upon Rural Regions Based on Multi-Resolution Image Analysis and Gabor Filters

Hang Jin[1], Marc Miska[1], Edward Chung[1], Maoxun Li[2] and Yanming Feng[3]
[1]Smart Transport Research Centre, Queensland University of Technology, Brisbane
[2]College of Urban Economics and Public Administration,
Capital University of Economics and Business, Beijing
[3]Faculty of Science and Technology, Queensland University of Technology
[1,3]Australia
[2]PR China

1. Introduction

Accurate, detailed and up-to-date road information is of special importance in geo-spatial databases as it is used in a variety of applications such as vehicle navigation, traffic management and advanced driver assistance systems (ADAS). The commercial road maps utilized for road navigation or the geographical information system (GIS) today are based on linear road centrelines represented in vector format with poly-lines (i.e., series of nodes and shape points, connected by segments), which present a serious lack of accuracy, contents, and completeness for their applicability at the sub-road level. For instance, the accuracy level of the present standard maps is around 5 to 20 meters. The roads/streets in the digital maps are represented as line segments rendered using different colours and widths. However, the widths of line segments do not necessarily represent the actual road widths accurately. Another problem with the existing road maps is that few precise sub-road details, such as lane markings and stop lines, are included, whereas such sub-road information is crucial for applications such as lane departure warning or lane-based vehicle navigation. Furthermore, the vast majority of road maps are modelled in 2D space, which means that some complex road scenes, such as overpasses and multi-level road systems, cannot be effectively represented. In addition, the lack of elevation information makes it infeasible to carry out applications such as driving simulation and 3D vehicle navigation.

Traditional methods for acquiring road information include i) ground surveying and ii) delineating roads from remotely sensed imagery (Zhang & Couloigner, 2004). Ground surveying can be carried out by using devices such as total stations and GPS receivers. As both devices are point-based, rendering this method labour-intensive and time-consuming, and therefore more suitable for detailed road surveying for small areas rather than for large-scale road mapping. Road information can be delineated from remote sensing images in three ways: i) manual delineation, ii) semi-automated extraction, iii) and fully automated detection. Manual extraction of roads from remotely sensed imagery is a simple stretching operation. However, the operation is impractically time consuming when the scenes are very complex. In addition, not only are such complex maps required for large geographic areas, frequent updating is also needed. In the semi-automatic road extraction method, approximations or seed points are given manually followed by an automatic algorithm which uses these approximations as input to enable them to automatically extract the road. Approximations can be a starting point, an ending point, intermediate points, road directions, road widths, and prior knowledge from a GIS database (Zhang, 2003). Full automatic road feature extraction is pursed by automating the selection of the necessary initial information.

As well as the advancement of innovative sensors and platforms, road network spatial information can be acquired from aerial and satellite imagery, synthetic aperture radar (SAR) imagery, airborne light detection and ranging (LiDAR) data, and from image sequences taken from ground-based mobile mapping systems (MMS) with different spatial and spectral resolutions (Quackenbush, 2004). Aerial images and LiDAR point clouds are promising data sources for generating road maps and updating available maps to support various activities and missions of government agencies and consumers (Mokhtarzade & Zoej, 2007). However, it has often been the case that while large amounts of high resolution aerial images and dense LiDAR data are being collected, piled up and remain unprocessed or unused, new data sets are continuously being gathered. This phenomenon is caused by the fact that development of automatic techniques for processing aerial imagery and LiDAR data is far behind that of the hardware sensor technologies. Object extraction for full exploitation of these data sources is very challenging. There are more challenges for automatic road information extraction in urban areas due to its much more complex circumstances.

Research on road feature extraction from aerial and satellite images can be traced back to the 1970s (Bajcsy & Tavakoli, 1976). Over three decades, a large number of automatic and semi-automatic algorithms have been attempted. Although many different approaches have been developed for the semi-automatic or automatic extraction of road information, none of these can solve all the problems without human interactions. This is because of the wide variations of roads (urban, rural, precipitous) and the complexities of their environment (occlusions caused by cars, trees, buildings, shadows etc.) (Poullis & You, 2010). It is worth noting that the existing road feature generation algorithms are all task-based and data-based. For instance, road surfaces have a quite different appearance from pavement markings; thus, approaches that are suitable for road surface extraction usually cannot be applied in the detection of pavement markings without modification. Due to the inherent difference in the data style, methods utilized for road extraction in aerial images may not be appropriate for LiDAR data sets. Therefore, in this work, an effective road information extraction system, which deals with road features in rural and urban regions respectively, is proposed based on very high resolution (VHR) aerial images.

The research is structured to present the main contributions as follows. Section 2 provides a review of the relevant work published over the past 20 years. Road feature extraction for rural and urban areas from high spatial resolution remotely sensed imagery is discussed separately in this section. In Section 3, an effective road network extraction method is presented. The homogeneity histogram thresholding algorithm utilized to detect road surface from VHR aerial images, and detected road features are then thinned and vectorized to reconstruct the digital road map. A novel road surface and lane marking extraction approach is presented in Section 4, which detects road surface from VHR aerial images based on support vector machine (SVM) classification method, and the lane markings are further generated using 2D anisotropic Gaussian filter as well as Otsu's thresholding algorithm. Concluding remarks and future work recommendations are given in Section 5.

2. Review of the related work

The review conducted by Mena (2003) cites more than 250 road extraction studies, and classifies different road extraction approaches based on three principal factors: i) the preset objective, ii) the extraction technique applied, and iii) the type of sensors utilized. Although the developed approaches exhibit a variety of methodologies and techniques, different categorizations for road extraction work can still be sought in order to better match the available data and methods to its ultimate purpose. In this review, we consider the use of major state-of-the-art data sources, aerial imagery, airborne LiDAR data, and categorize the existing road extraction methods into two classes, i) road detection in rural or non-urban regions, and ii) urban area road extraction. As the aerial imagery and LiDAR data are usually collected in the same flight missions, the extraction of road information from LiDAR data only is uncommon. This review is by no means exhaustive; instead, it focuses mainly on commonly used road extraction techniques.

Subsection 2.1 examines the work on rural area road extraction, and the review of road detection in urban regions is presented in Subsection 2.2. In addition, a brief summary of the road pavement marking extraction algorithms is provided in Subsection 2.3. Last but not least, the qualitative and quantitative evaluation of results is reviewed in Subsection 2.4.

2.1 Rural road extraction techniques

Roads in rural or non-urban areas have characteristics such as constant widths, continuous curvature changes, and homogeneous local orientation distributions, which can moderate the complexity of their extraction. Basically, rural road extraction approaches, either semi-automatic or automatic, can be classified into i) artificial intelligent, ii) multi-resolution analysis, iii) snakes, iv) classification, and v) template matching.

An automatic road verification approach based on digital aerial images as well as GIS data is developed in (Wiedemann & Mayer, 1996) as a part of the update procedure for GIS data. The candidates for roadsides, which are obtained by searching the surroundings of GIS road-axes in the image based on profiles, are tested, and a measure of confidence is also calculated. However, user interaction is still required, as the results of the method are far from perfect. Roads that do not exist in the GIS data will not be detected.

In (Doucette et al., 2001), a fully automated road extraction strategy based on Kohonen's self-organizing map (SOM) is proposed to detect road information in high-resolution multi-spectral aerial imagery. The core algorithms implemented include i) anti-parallel edge centerline extractor, ii) fuzzy organization of elongated regions, and iii) self-organizing road finder. A covariance-based principal component analysis (PCA) is performed to determine the intrinsic dimensions of the image bands, and to classify the image using a maximum likelihood classifier with manually selected training samples. The extraction results over several different areas and sensors show that the highest extraction quality and correctness rates are from anti-parallel edge analysis of spectral band and class layers, respectively.

Rellier et al. (2002) propose a model to locally register cartographic road networks on SPOT satellite image based on Markov random fields (MRF) so as to correct the errors and improve map accuracy. The method first translates the road network into a graph where the nodes are characteristic points of the roads. Then local registration is performed by defining a model in a Bayesian framework. One interesting point of the model is that the registration is done locally, which is very useful when the map exhibits local errors. The biggest problem with the model is still the computational time, which remains too long due to the frequency of computations of the path between nodes.

To extract roads from aerial images, Amo et al. (2006) employ the region competition algorithm, a mixed approach which combines region growing techniques with active contour models. Region growing makes the first step faster and region competition delivers more accurate results. However, this method is appropriate for roads in agricultural fields only, where roads are quite homogeneous and their homogeneity is sufficiently different from that of their surroundings.

Mayer et al. (1998) utilize the ribbon snake for the extraction of salient roads from aerial images based on the extracted lines at a coarse scale and the variation of road width at a fine scale. Non-salient roads are extracted by connecting two adjacent ends of salient roads with a road hypothesis, which is then verified based on homogeneity and the constancy of width. Finally, a closed snake is initialized inside the central area of the junction and expanded until delineating the junction borders. Mayer's method can overcome some problems such as extraction of shadowed and occluded roads, but it cannot deal with the complex road scenario in urban areas.

Laptive et al. (2000) use ribbon snakes to remove irrelevant structures extracted by a preliminary line detection algorithm at a coarse resolution. The method initializes a ribbon snake for each line detected and sets the width property to zero. The snake positions are optimized at a coarse scale to get a rough approximation of the road position. A second optimization process is used at a finer scale where the road position precision was increased and the width property expanded up to the structure boundary. Finally, road width thresholding is applied in order to discard any irrelevant structures.

A prior work for road detection based on image segmentation is conducted by Wang and Newkirk (1988), where a system is developed for automated highway network extraction from Landsat Thematic Mapper (TM) imagery supported by knowledge analysis and expert system. Three steps are involved in the system: i) binary image production, ii) tracing and feature extraction, and iii) highway identification. K-means clustering is employed to classify the image into two categories: road and non-road features. Analysis and processing are then performed on the linear patterns which are generated by labeling the binary image using a tracing algorithm. The proposed method is fairly simple and fully automatic, but the experiments are limited to the extraction of highways in rural areas.

Amini et al. (2002) utilize a segmentation method called the split and merge algorithm to automatically extract roadsides from large-scale image maps. The proposed method consists of two stages: i) straight lines extraction, and ii) roads skeleton extraction. The authors firstly generate a simpler image by grey scale morphological algorithms. Then the split and merge algorithm is applied on the simplified image, which is converted to a binary image. After that, the binary image map objects are labeled using the connected component analysis (CCA), and the skeletons of roads are extracted in the classified image by morphological operations. The roadsides are finally extracted by combining the skeleton of roads and the generated straight line segments.

Steger et al. (1995) propose a multi-resolution road extraction approach, where a different extraction method is utilized for each scale level. One method is applied on a fine scale with 25 cm GSD, while the other is applied at a lower resolution, which is reduced by a factor of eight. The larger scale method extracts roads based on a structural model matching technique, while the smaller scale method detects lines based on the image intensity level. Finally, the outputs are combined by selecting roads that are extracted at both levels.

An approach based on particle filtering is proposed in (Ye et al., 2006) to automatically extract roads from high resolution imagery. The road edges are extracted by the Canny detector, then the edge point distribution and the similarity of grey value are integrated into the particle filter to deal with complex scenes. To handle road appearance changes, the tracking algorithm is allowed to update the road model during temporally stable image observations.

Baumgartner et al. (1999) extract roads from multi-resolution images based on the work of Heipke (1995). In this paper, they emphasize the concept of "road model" comprising explicit knowledge about geometry, radiometry, topology, and context. They firstly segment the aerial image into global contexts (forest, rural and urban) to guide the extraction process in the various regions. In the coarse image, the line features are extracted using Steger's algorithm (1998). In the fine image, parallel edges are extracted and grouped into rectangles, which are then connected into the road segments. Finally, roads are generated through grouping road segments and closing gaps between them.

Dal-Poz et al. (2005) present an automatic method for road seed extraction from medium and high resolution images of rural scenes. The road-sides candidates are firstly detected by the Canny edge detector; the road objects are then built based on a set of rules constructed from a prior road knowledge. The rules used to identify and build road objects consist of anti-parallelism, parallelism and proximity, homogeneity, contrast, superposition, and fragmentation. Due to incompatibility with any road objects, road crossings cannot be extracted.

2.2 Road extraction in urban areas

Roads in urban areas have some unique characteristics absent in rural areas. There are often many shadows and occluded regions on road surfaces in urban areas due to the obstruction of tall buildings, vehicles, and trees. Furthermore, the contrast between roads and surrounding objects deteriorates significantly, since roads, side-walks, building roofs, and parking lots are usually constructed using similar materials, such as concrete and asphalt. Therefore, road extraction in urban areas cannot copy or enhance the methods and procedures which have been effective in the rural road extractions, such as the algorithms discussed above. Instead, it is necessary to develop an automatic system that can extract road information accurately as well as deal with the effects of background objects like cars, trees, or buildings. The key techniques used to reconstruct the urban road model include road tracking, segmentation and classification, mathematical morphology, and model based road extraction, which will be depicted in detail in the following paragraphs.

Shukla et al. (2002) applies a path-following method to extract road from high-resolution satellite imagery by initializing two points to indicate the road direction. Scale space and edge-detection techniques are used as pre-processing for segmentation and estimation of road width. The cost minimization technique is used to determine the road direction and generate the next seeds. This method performs better than the work of (Kim et al., 2002) because it can generate seeds in different directions at intersections. The limitations are that the algorithm may not work on roads on which shadows are cast.

Zhao et al. (2002) imposes a semi-automatic method by matching a rectangular road template with both road mask and road seeds to extract roads from IKONOS imagery. A road mask is the road pixels generated from maximum likelihood classification, and the road seeds can be generated by tracing the long edge of the road mask. The problem is all of the extracted road masks are not road area, and not all the extracted long edges are road edge; this results in misclassification.

Kim et al. (2004) initializes one seed point on the centerline of the road to determine the position of the reference template. The orientation of the road centerline, which is calculated with Burn's algorithms, guides the optimal target window. A least square template matching approach, which puts emphasis on the central part of the road, is utilized to determine the new location of the next road template. The limitations of this algorithm are i) that it cannot work with shadows, which may terminate the tracking process, ii) that the operator must select the initial seeds on road central lines, and iii) that one seed can be used to extract only one direction, leading to too many seeds when the scene is large and complex.

Hu et al. (2004) present a semi-automatic road extraction method based on a piecewise parabolic model with zero-order continuity, which is constructed by seed points placed by a human operator. Road extraction becomes a problem of estimating the unknown parameters for each piece of the parabola, which could be solved by least square template matching based on the deformable template and the constraint of the geometric model. In densely populated areas, where roads have sharp turns and orthogonal intersections, many seed points need to be located, which results in degrading the efficiency.

Shi and Zhu (2002) propose an approach to extract road network in urban areas from high-resolution satellite images. The basic procedures include binary image production by a threshold selection interactively, and a line segment match for road network processing. Binary image production is not automatic and the threshold parameter may change with the variation of image input, so it lacks a degree of automatic process and robustness, and further improvement is required. Grey-scale mathematical morphology is tested as one of the potential solutions in the proposed approach.

Haverkamp (2002) extracts road centerlines in urban areas from road segments and intersections based on size, eccentricity, length of the object and spatial relationships between neighboring intersections. A vegetation mask is derived from multi-spectral IKONOS imagery, and these objects are generated by grouping pixels with similar road directional information, based on texture analysis in a panchromatic IKONOS imagery. This method requires the predetermination of road width, which is tuned to detect roads with a specific level of contrast and a low along-road variance.

Two novel methods are developed in (Wang, 2004) to extract roads from high-resolution satellite images. One is a semi-automated road extraction method based on profile matching optimized by an auto-tuning Kalman filter, and the other is based on edge-aided multi-spectral classification. Experimental results from several aerial images show that both methods could accurately extract road networks from IKONOS and QuickBird satellite images, and could significantly eliminate the misclassification caused by small driveways, house roofs connected with the road networks, and extensive paved grounds.

Based on the fact that structural information obtained using mathematical morphological operators can provide complementary information to improve discrimination of different urban features that have a spectral overlap, Jin and Davis (2004) present applications of mathematical morphology for urban features extraction from high-resolution satellite imagery. To efficiently extract the road networks, directional morphological filtering is exploited to mask out those structures shorter than the distance of a typical city block. Directional top-hat operation is employed to mask out bright structures shorter than a city block. Similarly, dark structures shorter than a city block could be marked out by thresholding on the directional bottom-hat images.

Zhu et al. (2005) extract road network from 1-meter spatial resolution IKONOS satellite images based on the mathematical morphology and a line segment match method. The authors firstly generate the binary road image by adopting morphological leveling. Secondly, the coarse road network is detected using the proposed "Line Segment Match Method", which determines straight parallel line segments corresponding to roads. The holes are finally filled by using mathematical morphological operation. The proposed algorithm is based on the assumption that roads are a darker tone compared with the surrounding features, which may induce some problems in different situations.

Valero et al. (2010) propose a method for extracting roads in very high resolution (VHR) remotely sensed images, based on the assumption that roads are linear connected paths. Two advanced directional morphological operators, path opening and path closing, are utilized to extract structural pixel information; these remain flexible enough to fit rectilinear and slightly curved roads segments, due to their independence from the choice of a structural element shape. Morphological profiles are used to analyze object size and shape features so as to determine candidate roads in each level, since the morphological profiles of pixels on the roads are similar. Finally, a classical post-processing is employed to link the disconnected road segments using higher level representations (Tupin et al., 1998).

A Gibbs point process framework, which is able to simulate and detect thin networks from remotely sensed images, is constructed in (Stoica et al., 2004) to form a line-network for the road segments connection. The estimate for the network is found by minimizing an energy function. In order to avoid local minima, a simulated annealing algorithm based on a Monte Carlo Dynamics is utilized for finite point processes.

Based on Gaussian scale-space theory, a Gaussian comparison function is developed for extracting the linear road features from urban aerial remote sensing images (Peng & Jin, 2007). The curvilinear structures of the roads are verified, grouped and extracted, based on locally oriented energy in continuous scale-space combining the geometric and radiometric features. The system can significantly reduce computation complexity in the line tracking, and can effectively depress the zero drift caused by Gaussian smoothing, comparing with other edge-based line detection algorithms. The proposed curvilinear feature detection method is tested to be superior to the Canny operator and the Kovesi detector, in that it can detect not only urban highways but also the non-salient rural roads.

Peng et al. (2008) update digital road maps in dense urban areas by extracting the main road network from VHR QuickBird panchromatic images. A multi-scale statistical data model, which integrates the segmentation results from both coarse and fine resolution, is employed to overcome the difficulties caused by the complexity of information contained in VHR images. Furthermore, an outdated GIS digital map is utilized to provide specific prior knowledge of the road network. The experiments indicate that the combination of generic and specific prior knowledge is essential when working at full resolution.

2.3 Lane marking extraction techniques

The popular method for road pavement marking reconstruction is through a vehicle-based mobile mapping system (MMS), where the road lane markings can be detected and reconstructed in the field using laser scanners or close range photogrammetric imagery. Due to the difference in devices used and types of features fused, approaches developed for lane feature extraction have been quite distinct from one another. For instance, lane markings are extracted based on structures (Lai & Yung, 2000), image classification (Jeong & Nedevschi, 2005), and frequency analysis (Kreucher & Lakshmanan, 1999). An exhaustive review of road marking reconstruction approaches using MMS can be seen in (Soheilian, 2008). Although accurate lane features can be obtained through MMS, it is costly and time-consuming to produce lane data over large areas.

Lane information reconstruction through feature extraction from remote sensed images has been a long-standing research topic within the photogrammetry and remote sensing community. However, due to the limitation of the ground resolution of images, the majority of existing approaches concentrate on the detection of road centerline rather than sub-road details. Research efforts have been focused in a number of institutions, resulting in various approaches to the problem, including multi-scale approaches (Baumgartner et al., 1999), knowledge-based extraction (Trinder & Wang, 1998) and context cues (Hinz & Baumgartner, 2000).

Only a few approaches involve the detection of lane marking in road extraction. Steger et al. (1997) extract the collinear road markings as bright objects with the algorithm given in (Steger, 1996) in large scale photographs when the roadsides exhibit no visible edges. Only the graph search strategy is adapted to extract road markings automatically, and a best-first search from a few salient road markings is also utilized. The strategy adds the road marking to the best connection evaluation only, which would add a global evaluation step following each marking, and try to add a new road marking if the directions of the road markings are not extracted perfectly.

In a more recent work, Kim et al. (2006) build a system to extract pavement information in complex urban areas relying on a set of simple image processing algorithms. The pavement information included land and symbol markings that guide direction, and the geometric properties of the pavement markings and their spatial relationships are analyzed. Moreover, road construction manuals and a series of cutting-edge algorithms, including template matching, are involved in the analysis. The evaluation of accuracy by comparing the data with manually plotted ground truth data validate that road information can be extracted efficiently to an extent in a complex urban region.

Tournaire et al. (2009) propose a specific approach for dashed lines and zebra crossing reconstruction. This approach relies on external knowledge introduced in the detection and reconstruction process, and is based on primitives extracted in the images. The core of the approach relies on defining geometric, radiometric and relational models for dashed lines objects. The model also deals with the interactions between the different objects making up a line, which means that the algorithm introduces external knowledge taken from specifications. To sample the energy function, the authors also use Green's algorithm, complete with a simulated annealing, to find its minimum.

2.4 Result evaluations

Internal diagnosis and external evaluation for the extracted road models are two important aspects of assessment of the relevant automatic road extraction system (Wiedemann et al., 1998). However, relatively little work has been carried out in this area.

In (Heipke et al., 1997) and (Wiedemann et al., 1998), an external evaluation approach of automatic road extraction algorithms is developed by comparison of these to manually plotted linear road axes used as reference data. The quality measures proposed for the automatically extracted road data comprise completeness, correctness, quality, redundancy, planimetric RMS differences, and gap statistics, and are all aimed at exhaustive evaluation as well as assessing geometrical accuracy. The proposed evaluation method is tested by comparing evaluations of three different automatic road extraction approaches, and demonstrating its applicability.

An in-depth usability evaluation of a semi-automated road extraction system is presented in (Wilson et al., 2004), highlighting both strengths and areas for improvement. The evaluation is principally conducted on the timing and statistical analysis as well as on factors that affect the extraction speed. Peteri et al. (2004) present a method to guide the determination of a reference based on statistical measures from several image interpretations. A tolerance zone representative of the variations in interpretation is defined that allows both the determination of the uncertainty of the reference object and the possibility of defining criteria for a quantitative evaluation. A few criteria defined by Musso and Vuchic (1988), including the size, form, and topology indices of the road network, are employed to carry out evaluation of the planimetric accuracy and the spatial characterization of a road network.

To qualitatively evaluate the performance of the semi-automatic road extraction algorithms, four criteria (correctness, completeness, efficiency, and accuracy) are utilized in (Zhou et al., 2006) and further in (Zhou et al., 2007). Completeness and correctness are the priority criteria in cartography, while the efficiency measurement principally takes the savings of human input into consideration. Tracking accuracy is assessed as the root mean square error between the road tracker and the human input.

To sum up, the typical result evaluation approach for road extraction has been carried out by comparing the generated roads with manually plotted reference data. Correctness and completeness are the two most frequently used criteria, while other measurements are dependent on specific road extraction algorithms and objectives.

3. Road extraction in rural regions

In this section, we developed a new approach for automatic road network extraction, where both spatial and spectral information from aerial photographs or pan-sharpened QuickBrid images is systematically considered and fully used. The proposed approach is performed by the following three main steps: (i) the image is classified based on homogeneity histogram segmentation to roughly identify the road network profiles; (ii) the morphological opening and closing is employed to fill tiny holes and filter out small road branches; and (iii) the extracted road surface is further thinned by a thinning approach, pruned by a proposed method and finally simplified with Douglas-Peucker algorithm.

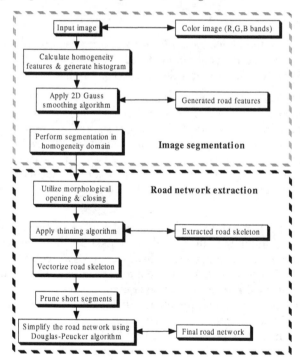

Fig. 1. Flowchart of the proposed method

As a popular technique for image segmentation, histogram based thresholding only takes the occurrence of the gray level into account without any local information. But the segmentation based on the property of image homogeneity involves both the occurrence of the gray levels and the neighbouring homogeneity value among pixels; thus it will be employed in this study to obtain a more homogeneous segmentation result. Gaussian smoothing algorithm is then applied to this obtained homogeneity histogram, which can, in turn, ease the threshold finding procedure for segmentation. After achieving image segmentation, morphological opening and closing is utilized to remove small holes and noise from the road surface as well as narrow pathways connected to the main road. Then a thinning method is further applied to extract the skeleton of the road network. Finally, the generated road network is vectorized, and then pruned and simplified respectively by a proposed pruning method and Douglas-Peucker algorithm. Fig. 1 illustrates the flowchart for the developed approach. Basically, the performance includes two individual processes, namely, image segmentation and road network extraction, which will be elaborated in the following sections.

3.1 Image segmentation

Road network is detected using homogeneity histogram segmentation, which comprises the following two basic operations: contrast stretching, homogeneity histogram construction and smoothing.

Contrast stretching

Colour images can be represented by linear RGB colour space or their non-linear transformation of RGB, e.g. HSI (hue, saturation and intensity). It is, in general, easier to discriminate highlights and shadows in a colour image by using the HSI colour space than the RGB colour space, but the hue is rather unstable at low saturation and makes the segmentation unreliable. Although the three basic RGB components are highly correlated in RGB colour space, the latter is applied in this paper due to its efficiency in distinguishing small variations in colour.

All of the RGB channels, especially the blue channel, in an original aerial photo (Fig. 2 (a)) have relative contrast deficiency which will impose challenges to the segmentation process. Therefore, contrast stretching is individually applied to each channel by assigning 5% and 95% in the histogram as the lower and upper bounds over which the image is to be normalized. It is clear that the contrast stretched images (shown in figure 2 (b), (c) and (d))have significantly higher contrast than the original RGB channels.

Homogeneity histogram construction

A general concept of the homogeneity histogram is referred to Cheng (2000). The homogeneity histogram takes into account not only the gray level but also spatial information of pixels with respect to each other. Therefore, homogeneity histogram thresholding tends to be more effective in finding homogeneous regions than histogram thresholding approaches.

The homogeneity vector of the pixel with its eight neighbours is calculated by Z-function, allowing the homogeneity histogram to be defined by normalization of the homogeneity vector. The normalized homogeneity histogram for Red, Green and Blue channels are shown in Fig. 3.

It is still difficult to detect the modes of homogeneity histogram in the above normalized homogeneity histogram when they are corrupted by noise. Therefore, once the homogeneity histogram for R, G and B channels are established, Gaussian filter is firstly applied to smooth them, instead of finding the thresholds directly by a complex peak finding algorithm proposed by Cheng (2000). In Gaussian filtering process, the spread parameter σ, which determines the

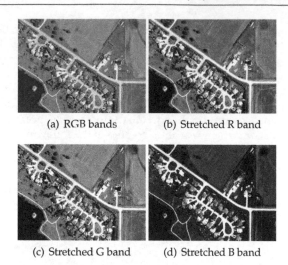

(a) RGB bands   (b) Stretched R band

(c) Stretched G band   (d) Stretched B band

Fig. 2. The original aerial photo and its Red, Green, and Blue channels.

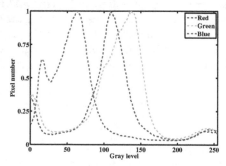

Fig. 3. Normalized homogeneity histogram for Red, Green and Blue channel images

amount of smoothing, is determined with the algorithm proposed by Lin et al. (1996). Each peak in the homogeneity histogram represents a unique region. Accordingly, the valleys in the homogeneity histogram can be used as the thresholds for segmentation, as they can be easily found in the smoothed homogeneity histogram (see Fig. 4).

Each colour channel is segmented using the above obtained thresholds separately, and then all three segmented channel images are fused to yield the final result of segmentation (see e.g., Fig. 5). It is observed from Fig. 5 (d) that almost all the road networks are correctly extracted, but there are still many small family driveways connected to road networks and many house roofs are misclassified into the road network. These make it impossible to obtain an accurate road network without further processing.

3.2 Road network extraction

Up until now we have obtained the segmented result for road objects (see e.g. Fig. 6(a)), but the probability of misclassification is still relatively high and many small holes enclose the main road network. These holes and pathways must be removed to correctly extract the road

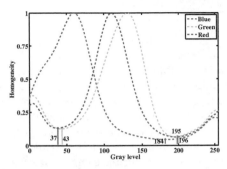

Fig. 4. Normalized homogeneity histogram for Red, Green and Blue channel images

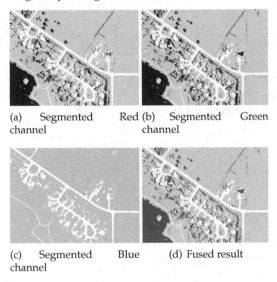

(a) Segmented Red channel

(b) Segmented Green channel

(c) Segmented Blue channel

(d) Fused result

Fig. 5. The segmented Red, Green and Blue channel images, and the final fused result.

skeleton. In this section, a novel road network extraction approach is developed to accurately extract road networks from a segmented road image. This extraction process includes two main steps: morphological operation and thinning and vectorization.

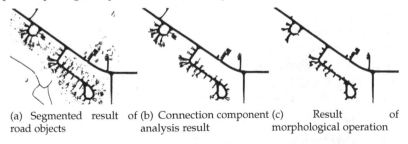

(a) Segmented result of road objects

(b) Connection component analysis result

(c) Result of morphological operation

Fig. 6. The noise removal of the segmented result.

Morphological operation

Mathematical morphology is a structure-based mathematical set theory that uses set operations such as union, intersection and complementation, so it is favoured for high-resolution image processing (Mohammadzadeh et al., 2006). Connected component analysis is firstly used to group pixels into different components based on pixel connectivity, then components whose surface area are smaller than a given threshold will be removed. The filtered image is shown in Fig. 6 (b), it can be clearly seen that all the misclassified objects unconnected to the main road network were removed. Morphological closing is then applied to remove small holes and noise from the road surface, while an opening operation is used to eliminate small pathways with a structuring element size that is smaller than the main roadâĂŹs width but larger than those of the pathways, resulting in the extracted road network as shown in Fig. 6 (c).

Thinning and vectorization

After the morphological operation, we further employ the thinning algorithm proposed by Wang and Zhang (1989) to extract the road skeleton, where the real road is replaced by its centreline with representation by a pixel. To remove short dangling branches of the centrelines caused by driveways, a novel pruning algorithm is performed as follows.

N[1]	N[1]	N[1]
N[1]	P	N[1]
N[1]	N[1]	N[1]

Fig. 7. Pixel P and its eight neighbours.

First of all, we introduce the definitions of four-neighbourhood and eight-neighbourhood neighbour for point P in Fig. 7. Here four-neighbourhood refers to N[1], N[3], N[5] and N[7], while eight-neighbourhood neighbour involves N[0], N[2], N[4] and N[6].

The pruning algorithm includes three steps:

Step 1 Find all the intersection points

1. Scan the image (up to bottom, left to right), if current pixel P has more than three foreground neighbours, namely, $\{N[x_i] \mid i = 1, 2, \cdots, k; k \geq 3, x_k = 0, 1, \cdots 7\}$, go to 2.
2. Initialize the feature point counter c=0, and then from i=1 to k, set c=c+1 if either condition (a) or (b) is satisfied.
   (a) $N[x_i]$ is four-neighbourhood neighbour of P.
   (b) $N[x_i]$ is eight-neighbourhood neighbour of P and neither $N[x_i - 1]$ nor $N[x_i + 1]$ is foreground pixel. P is supposed to be a intersection point if $c \geq 3$.

Step 2 Line tracking

1. If there is no intersection point in the image, then go to 3.
2. Tracking lines from the intersection point.

Road Feature Extraction from High Resolution Aerial Images Upon Rural Regions Based on Multi-Resolution
Image Analysis and Gabor Filters

183

    (a) Start from the intersection point P found in Step 1, initialize n (number of P's feature points) arrays to store lines started from P.

    (b) Set the current tracking pixel to background after storing its position into the array, go on using the condition in Step 1 to find the next pixel on current tracking line until moving to the endpoint or other intersection point.

  3. Tracking lines from endpoint.

    (a) Scan the image (up to bottom, left to right).

    (b) Find the endpoint, start line tracking from it and set the pixels on the line to background (endpoint's number of feature point is 1 using the condition in Step 1).

    (c) Go on scanning until to the end of the image.

Step 3 Small line pruning

  1. Delete line from the line array if both the following conditions are satisfied:

    (a) The length of line is shorter than the threshold T.

    (b) If both endpoints of the line are not intersection points, and then go to Step 1.

  2. Output the final result.

Finally, Douglas-Peucker simplification algorithm, which not only decreases the number of data points but also retains the similarity of the simplified shape to the original one as close as possible, is employed to the pruned line network. The whole procedure of vectorization and simplification is shown in Fig. 8. The vectorization process consists of two steps: intersection point searching and line tracking, followed by small lines pruning and simplification. The final result is shown in Fig. 9. It can be seen that this approach works quite well that all the small road branches are removed.

3.3 Experimental results and evaluation

In order to demonstrate the efficient performance of the proposed procedures outlined in this paper, two additional experiments have been implemented from the QuickBird satellite images, and their extraction accuracies are also evaluated. The final road network extracted using the proposed method is shown in Figure 10. Almost all the main roads are correctly extracted. However, the developed method is still experiencing difficulties in road extraction from the images where indistinct contrast between the road surface and its surroundings, as well as shadows, exist. This is another important research topic to be resolved.

Variables	Completeness	Correctness	Quality
Figure 9	98.5%	96.2%	94.7%
Figure 10 (a)	98.8%	99.3%	98.1%
Figure 10 (a)	81.9%	98.2%	80.7%
Means	93.1%	97.9%	91.2%

Table 1. Evaluation of the test results.

Basing on the method developed by Wiedemann (1996) for evaluating automatic road extraction systems, we use three indexes to assess the quality of the generated road network. The completeness is defined as the percentage of the correctly extracted data over the reference data and the correctness represents the ratio of correctly extracted road data. The quality

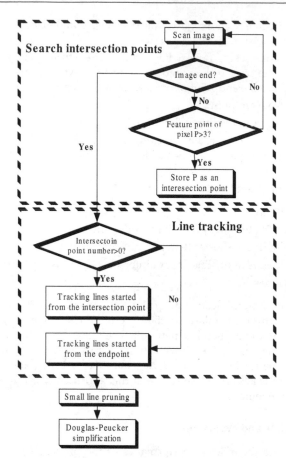

Fig. 8. Flowchart for implementation of the vectorization and pruning.

Fig. 9. Final centreline laying on the original road surface.

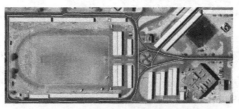

(a) Riyadh, Saudi Arabia

(b) Hurghada, Egypt

Fig. 10. Road extraction tests on QuickBird images.

is a more general measure of the final result combining the completeness and correctness. The optimum values for the above three defined indexes are all equal to one. Comparing automatically achieved results from the proposed process with the manual ones, the following quantified indicators have been calculated and presented in Table 1. The results demonstrate that the proposed method achieved a significantly high level of accuracy.

3.4 Summary

In this section, we have presented a new approach for road extraction from large scale remote sensing images. The tests have demonstrated that considerable success can be achieved by adopting the overall flowchart presented in this paper, particularly when the contrast between road surface and background is distinct, and there is a significant proportion of road surface in the image. Importantly, a novel algorithm is developed to vectorize and prune the extracted road network. The experimental results for road extraction from aerial photo and QuickBird satellite images demonstrate that the proposed approach could extract most of the main roads despite the fact that some roads are missing or are slightly distorted.

4. Road detection in urban areas

Accurate and detailed road models are of great importance in many applications, such as traffic monitoring and advanced driver assistance systems. However, the majority of road feature extraction approaches have only focused on the detection of road centerline rather than the lane details. Only a few approaches involved the detection of lane markings in the road extraction. For instance, Steger et al. (1997), Hinz and Baumgartner (2003), and Zhang (2004) extracted the road markings in their attempts to obtain clues as to the presence of road surface. Consequently, important requirements (Tournaire & Paparoditis, 2009) such as robustness, quality, completeness, are achieved less consistently compared to the lane level applications. In more recent works, Kim et al. (2006) and Tournaire et al. (2009) presented systems for pavement information extraction from remote sensing images with high spatial resolution.

In this section, the support vector machine (SVM) and Gabor filters are introduced into a framework for precise road model reconstruction from aerial imagery. The experimental practices using a data set of aerial images acquired in Brisbane, Queensland are utilized to evaluate the effectiveness of the proposed strategy.

4.1 Methodology

Supervised SVM image classification technique is employed to segment the road surface from other ground details, and the road pavement markings are detected on the generated road surface with Gabor filters.

An SVM is basically a linear learning machine based on the principal of optimal separation of classes (Vapnik, 1998). The goal is to find a linear separating hyperplane that separates the classes of interest provided the data is linearly separable. The hyperplane is a plane in a multidimensional space and is also called a decision surface or an optimal separating hyperplane or a maximal margin hyperplane.

Consider a set of l labelled training patterns $(x_1, y_1), (x_2, y_2), \cdots, (x_i, y_i), \cdots, (x_l, y_l)$, where x_i denotes the i-th training sample and $y_i \in \{1, -1\}$ denotes the class label. If the data are not linearly separable in the input space, a non-linear transformation function $\Phi(\cdot)$ is used to project x_i from the input space to a higher dimensional feature space. An optimal separating hyperplane is constructed in the feature space by maximizing the margin between the closest points $\Phi(x_i)$ of two classes. The inner-product between two projections is defined by a kernel function $K(x, y) = \Phi(x) \cdot \Phi(y)$. The commonly used kernels include polynomial, Gaussian RBF, and Sigmoid kernels. Further details about kernels can be found in (Vapnik, 1998).

The decision function of the SVM is defined as

$$f(x) = w \cdot \Phi(x) + b = \sum_{i=1}^{l} \alpha_i y_i K(x, x_i) + b$$

subject to $\sum_{i=1}^{l} \alpha_i y_i = 0$ and $0 \leq \alpha \leq C$, where C denotes a positive value determining the constraint violation during the training process.

Due to its properties of non-parametric, sparsity, and intrinsic feature reduction, SVM is superior to conventional classifiers, such as the maximum likelihood classifier, for image classification in very high resolution (VHR) remotely sensed data, since the estimated distribution function usually employs the normal distribution, which may not represent the actual distribution of the data (Huang & Zhang, 2008).

4.1.1 Gabor filters

2D Gabor filters, extended from 1D Gabor by Daugman (1985), have been successfully applied to a variety of image processing and pattern recognition problems, such as texture analysis, and image segmentation. 2D Gabor filters can be used to extract the road lane markings thanks to their following properties: (i) tuneable to specific orientations, (ii) adjustable orientation bandwidth, and (iii) robust to noise. Furthermore, it has optimal joint localization in both spatial and frequency domains. Therefore, Gabor filters can be considered as orientation and scale tunable edge and line (bar) detectors (Manjunath & Ma, 1998), which makes these a superior tool to detect the geometrically restricted linear features, such as road pavement markings.

Gabor functions

The general functionality of the 2D Gabor filter family can be represented as a Gaussian function modulated by a complex sinusoidal signal. Specifically, the 2D Gabor filter can be defined in both the spatial domain $g(x, y)$ and the frequency domain $G(u, v)$. The 2D Gabor function in spatial domain can be formulated as (Cai & Liu, 2000):

$$g(x,y) = \exp\left\{-\pi\left(\frac{x_r^2}{\sigma_x^2} + \frac{y_r^2}{\sigma_y^2}\right)\right\} \exp\left\{j2\pi(u_0 x + v_0 y)\right\}$$

Its 2D Fourier transform is expressed as

$$G(u,v) = \exp\left\{-\pi\left[(u - u_0)_r^2 \sigma_x^2 + (v - v_0)_r^2 \sigma_y^2\right]\right\}$$

where $j = \sqrt{-1}$; (x_0, y_0) indicates the peak of the Gaussian envelope; (σ_x, σ_y) are the two axis scaling parameters of the Gaussian envelope; (u_0, v_0) presents the spatial frequencies of the sinusoid carrier in Cartesian coordinates, which can also be expressed in polar coordinates as (f, ϕ), where $f = \sqrt{u_0^2 + v_0^2}$, $\phi = \arctan(v_0 / u_0)$, and the subscript r stands for a rotation operation as follows:

$$x_r = x \cos\theta + x \sin\theta$$
$$y_r = -x \sin\theta + y \cos\theta$$

where θ is the rotation angle of the Gaussian envelope.

Determination of Gabor filter parameters

Road markings, which are presented as linear features with certain widths and orientations within local areas, can be considered as rectangular pulse lines. The correct determination of Gabor filter parameters is the central issue for lane pavement markings' extraction process. In order to effectively and accurately extract road lane markings with different sizes and thicknesses from aerial images using Gabor filters, we proposed an efficient method to determine the Gabor filter parameters.

Determination of θ

θ stands for the orientation of the span-limited sinusoidal grating. The orientation θ ($\theta \in [0, \pi)$) of Gaussian envelope is given as perpendicular to the direction φ ($\varphi \in [0, \pi)$) of the road surface by:

$$\theta = (\varphi + \pi/2) \,\%\, \pi$$

where % is the modulo operator.

Determination of f

f is the frequency of the sinusoid, which determines the 2D spectral centroid positions of the Gabor filter. This parameter is derived with respect to the width of road lane markings. In order to produce a single peak for the given lane line as well as discard other ground objects, such as white vehicles, the frequency f of the Gabor filter must satisfy the following conditions:

$$1/W' < f \le 1/W_m$$

where where W_m is the width of the road marking in pixel, and W' is the width of other white features. The details of the proofing process can be referred to (Liu et al., 2003).

In our experiments, we set $f = 1/W_m$, which will produce only a single peak in the output of the filter on road markings regardless of the values of σ_x and σ_y.

Determination of σ_x and σ_y.

The parameters σ_x and σ_y determine the spread of the Gabor filter in Ï̈T and Ï̈ÿ directions respectively. According to (Liu et al., 2003), σ_x and σ_y have the following parameter constraint:

$$\sigma_y = k\sigma_x$$

where k is a constant. As the road lane markings have strict orientation and enough distance between adjacent lanes, we set k=1 to simplify the calculation.

The relationship between the orientation bandwidth $\triangle\theta$ and the frequency f within the frequency domain is illustrated in figure 1, which can be given by:

$$\triangle\theta = 2\arctan\left(\frac{l}{f}\right)$$

where $\triangle\theta$ is the orientation bandwidth. It give:

$$l = f\tan(\triangle\theta/2)$$

Applying the 3dB frequency bandwidth in V direction when $\phi = 90°$ to equation (2), we have

$$G(u_0, h)\,|_{\phi=90} = \exp\left[-\pi(h\sigma_x)^2\right] = \sqrt{2}/2$$

It gives

$$\sigma_x = \frac{\sqrt{\dfrac{\ln 2}{2\pi}}}{d \tan\left(\triangle\theta/2\right)}$$

According to orientation bandwidths of cat cortical simple cells (Liu et al., 2003), the mean angle covers a range from 26° to 39°. After examining the line extraction results over the above range, we find it appropriate to set $\triangle\theta = 30°$. Then σ_x and σ_y can be further obtained by:

$$\sigma_x = \sigma_y = 0.58/f$$

4.2 Experiments and discussion

The objective of the experiment is to determine the performance of the proposed road feature extraction approach quantitatively over the study area. A dataset of aerial images located in South Brisbane, Queensland have been selected as the study areas. The selected aerial images consist of three bands: Red, Blue and Green, with Ground Sampling Distance (GSD) of 7 cm. Fig. 11 shows one of the testing images.

Fig. 11. One testing site (4096×4096 pixels).

Several training samples were used to train the support vector machine and the resulting model was used to classify the whole image into two features: road and non-road. For the implementation of SVM, the software package LIBSVM by Chang and Lin (2003) was adapted. Gaussian RBF was used as the kernel function, and the constraint violation C was set to be 10. After the image classification, the connected component analysis was used to remove small noises misclassified into road class.

To this point, the road surface has been obtained using SVM classification. Gabor filter was then utilized to extract the lane marking features while restrain the affection from other ground objects. To reduce the calculation complexity, Principle Component Analysis (PCA) was applied on the color image and only the 1st component was chosen for Gabor filtering. The parameters of Gabor filters are determined as outlined in the previous section. For instance, the orientation of the lane markings shown in Fig. 11 is approximately 130 degrees. The average of width of the road markings is 6 pixels, thus the frequency f is set to be 0.17, while the axis scaling parameters σ_x and σ_y of the Gaussian function is set to be 3.4. The

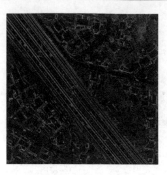

Fig. 12. Gabor filtered result.

filtered image is as illustrated in figure 4, which was then masked by the road surface acquired in the previous step.

Finally, the Gabor filtered image was then segmented by Otsu's thresholding algorithm, and directional morphological opening and closing algorithms were utilized to remove misclassified features. Some white linear features such as house roof ridges may be misclassified into lane markings, so we further utilized the extracted road surface in the previous step as a mask to remove these kinds of objects. The lane segments may also be corrupted by many facts: occlusion, e.g. trees above the road surfaces; worn-out painting of lane lines; dirty markings on the road surfaces. We eliminated the affection from vehicles in the road markings extraction by utilizing the following two indicators: (i) elongation - the ratio of the major axis to the minor axis of the polygon, and (ii) lengths of the major and minor axis. The elongation measure of vehicle is smaller than the road lane markings, and the length of the major and minor axis of vehicle are within certain ranges. In this experiment, the major axis length of the vehicle is set to be within 2 to 10m, while minor axis is set to be between 1.5m and 3m. The extracted pavement markings are superimposed on the road surfaces, as given in Fig. 13.

Fig. 13. Final result.

The quantitative evaluation of the experimental results is achieved by comparing the automated (derived) results against a manually compiled, high quality reference model. Following the concept of error matrix, the evaluation matrices for the accuracy assessment of road surfaces detection can be defined at the pixel level as follows:

Road Feature Extraction from High Resolution Aerial Images Upon Rural Regions Based on Multi-Resolution
Image Analysis and Gabor Filters

191

1. Detection rate

$$d = \frac{TP}{TP + FN}$$

2. False alarm rate

$$f = \frac{FP}{TP + FP}$$

3. Quality

$$q = \frac{TP}{TP + FP + FN}$$

In the above equation, TP (true positive) is the number of road surface pixels correctly identified, FN (false negative) is the number of road surface pixels identified as other objects, FP (false positive) is the number of non-road pixels identified as road surfaces.

The evaluation of the extracted pavement marking accuracy is carried out by comparing the extracted pavement markings with manually plotted road markings used as reference data as presented in (Wiedemann et al., 1998), and both data sets are given in vector representation. The buffer width is predefined to be the average width of the road markings, and we set it to be 15 cm in our experiment. Then the accuracy measures are given as:

1. Detection rate

$$d = \frac{length\,of\,the\,matched\,reference}{length\,of\,reference}$$

2. False alarm rate

$$f = \frac{length\,of\,the\,unmatched\,extraction}{length\,of\,extraction}$$

3. Quality

$$q = \frac{length\,of\,the\,matched\,reference}{length\,of\,extraction + unmatched\,reference}$$

Road boundaries and road markings are firstly digitized from the test images and used as ground truth. Three measures of the extraction results for both road surfaces and pavement markings are given in Table 2.

Test image	Road features	Detection rate	False alarm rate	Quality
Image I	Surface	91.8%	12.9%	80.3%
	Markings	93.3%	10.6%	83.7%
Image II	Surface	93.2%	7.2%	88.5%
	Markings	94.5%	2.7%	92.7%
Image III	Surface	88.3%	2.2%	86.2%
	Markings	83.5%	15.2%	71.8%

Table 2. Evaluation of the test results.

For the entire four test sites, nearly 90% of the road surfaces are correctly detected, and the relevant false alarm rate is about 10%. The completeness of road pavement marking extraction reaches above 87%, except for test site IV, which is seriously affected by shadows. The shadows on the road surfaces can reduce the intensity contrast between pavement markings and the road surface background, which makes it difficult to enhance the road markings using the Gabor filter. The average false alarm rate of the four test sites is about 10%.

4.3 Summary

In this section, an automatic road surface and pavement marking extraction approach from aerial images with high spatial resolution is proposed. The developed method, which is based on SVM image classification as well as Gabor filtering, can generate accurate lane level digital road maps automatically. The experimental results using the aerial image dataset with ground resolution of 7 cm have demonstrated that the proposed method works satisfactorily. Further work will concentrate on the process of seriously curved road surface and large images, which may be achieved by using knowledge based image analysis and image partition technique.

5. Conclusions and future work

5.1 Conclusions

In conclusion, we have presented an integrated approach for road feature extraction from both rural and urban areas. Road surface and lane markings have been extracted from very high resolution (VHR) aerial images in rural areas based on homogeneity histogram thresholding and Gabor filters. The homogeneity histogram image segmentation method takes into account not only the color information but also the spatial relation among pixels to explore the features of an image. We further proposed a road network vectorization and pruning algorithm, which can effectively eliminate the short tracks segments. In the urban area, the road surface is firstly classified by SVM image segmentation method, and then Gabor filter is further employed to enhance the road lane markings whilst constraining the effects of other ground features. The experimental results from several VHR satellite images in rural areas have indicated that over 95% of road networks have been correctly extracted. The omission of road feature is a result of occlusions, poor contrast with the surrounding scenario, and partial shadows over the road. This has preliminarily demonstrated that the presented extraction strategy for road feature extraction in rural areas is promising. Experiments with three typical test sites in urban areas have resulted in over 90% of the road surfaces being corrected extracted, with the misclassification rate below 10%. The correction rate for lane marking extraction is approximate 95%, and only about 10% of the other ground objects are misclassified as lane marking.

5.2 Future work

Although the proposed approach has generated satisfactory results on the testing datasets, problems still exist: for example, lane markings obstructed by vehicles may not be effectively detected. Therefore, future work will focus on the improvement of detection accuracy and precise model reconstruction. For instance, an automatic vehicle detection approach may be introduced to efficiently detect and remove vehicles from the road surface. GPS real-time kinematic positioning solutions from a probe vehicle could beappropriate for the recovery of lane markings in areas where there are large obstructions: for example, a large number of skyscrapers or trees would greatly deteriorate the extraction result in urban or forest areas. We also consider using the linear feature linking technique to connect the broken road features.

6. References

Amini, J., Saradjian, M. R., Blais, J. A. R., Lucas, C. & Azizi, A. (2002). Automatic road-side extraction from large scale imagemaps, *International Journal of Applied Earth Observation and Geoinformation* 4(2): 95–107.

Amo, M., Martinez, F. & Torre, M. (2006). Road extraction from aerial images using a region competition algorithm, *IEEE Transactions on Image Processing* 15(5): 1192–1201.

Bajcsy, R. & Tavakoli, M. (1976). Computer recognition of roads from satellite pictures, *IEEE Transactions on Systems, Man and Cybernetics* 6(9): 623–637.

Baumgartner, A., Steger, C., Mayer, H., Eckstein, W. & Ebner, H. (1999). Automatic road extraction based on multi-scale, grouping, and context, *Photogrammetric Engineering & Remote Sensing* 65(7): 777–785.

Cai, J. & Liu, Z. (2000). Off-line unconstrained handwritten word recognition, *International Journal of Pattern Recognition and Artificial Intelligence* 14(3): 259–280.

Chang, C.-C. & Lin, C.-J. (2003). Libsvm: a library for support vector machines, *Technical report*, Deptartment of Computer Science and Information Engineering, National Taiwan University.

Cheng, H. D. & Sun, Y. (2000). A hierarchical approach to color image segmentation using homogeneity, *IEEE Transactions on Image Processing* 9(12): 2071–2082.

Dal-Poz, A. P., Vale, G. M. D. & Zanin, R. B. (2005). Automatic extraction of road seeds from high-resolution aerial images, *Annals of the Brazilian Academy of Sciences* 77(3): 509–520.

Daugman, J. G. (1985). Uncertainty relation for resolution in space, spatial frequency, and orientation optimized by two-dimensional visual cortical filters, *Journal of Optical Society of America, A: Optics and Image Science* 2(7): 1160–1169.

Doucette, P., Agouris, P., Stefanidis, A. & Musavi, M. (2001). Self-organised clustering for road extraction in classified imagery, *ISPRS Journal of Photogrammetry and Remote Sensing* 55(5-6): 347–358.

Haverkamp, D. (2002). Extracting straight road structure in urban environments using ikonos satellite imagery, *Optical Engineering* 41(2107): 2107–2110.

Heipke, C., Mayer, H., Wiedemann, C. & Jamet, O. (1997). Evaluation of automatic road extraction, *International Archives of Photogrammetry and Remote Sensing* 32(3-2W3): 47–56.

Heipke, C., Steger, C. & Multhammer, R. (1995). A hierarchical approach to automatic road extraction from aerial imagery, *Society of Photographic Instrumentation Engineers (SPIE)* 2486: 222–231.

Hinz, S. & Baumgartner, A. (2000). Road extraction in urban areas supported by context objects, *International Archives of Photogrammetry and Remote Sensing* 33(B3/1): 405–412.

Hinz, S. & Baumgartner, A. (2003). Automatic extraction of urban road networks from multi-view aerial imagery, *ISPRS Journal of Photogrammetry and Remote Sensing* 58(1-2): 83–98.

Hu, X., Zhang, Z. & Tao, C. V. (2004). A robust method for semi-automatic extraction of road centerlines using a piecewise parabolic model and least squares template matching, *Photogrammetric Engineering & Remote Sensing* 70(12): 1393–1398.

Huang, X. & Zhang, L. (2008). An adaptive mean-shift analysis approach for object extraction and classification from urban hyperspectral imagery, *IEEE Transactions on Geoscience and Remote Sensing* 46(12): 4173–4185.

Jeong, P. & Nedevschi, S. (2005). Efficient and robust classification method using combined feature vector for lane detection, *IEEE Transactions on Circuits and Systems for Video Technology* 15(4): 528– 537.

Jin, X. & Davis, C. H. (2004). New applications for mathematical morphology in urban feature extraction from high-resolution satellite imagery, *Proceedings of the SPIE* 5558: 137–148.

Kim, J. G., Han, D. Y., Yu, K. Y., Kim, Y. I. & Rhee, S. M. (2006). Efficient extraction of road information for car navigation applications using road pavement markings obtained from aerial images, *Canadian Journal of Civil Engineering* 33(10): 1320–1331.

Kim, T., Park, S. R., Jeong, S. & Kim, K. O. (2002). Semi automatic tracking of road centerlines from high resolution remote sensing data, *the 23rd Asian Conference on Remote Sensing*, Kathmandu, Nepal.

Kim, T., Park, S.-R., Kim, M.-G., Jeong, S. & Kim, K.-O. (2004). Tracking road centerlines from high resolution remote sensing images by least squares correlation matching, *Photogrammetric Engineering & Remote Sensing* 70(12): 1417–1422.

Kreucher, C. & Lakshmanan, S. (1999). Lana: a lane extraction algorithm that uses frequency domainfeatures, *IEEE Transactions on Robotics and Automation* 15(2): 343–350.

Lai, A. & Yung, N. (2000). Lane detection by orientation and length discrimination, *IEEE Transactions on Systems, Man, and Cybernetics Part B: Cybernetics* 30(4): 539–548.

Laptev, I., Mayer, H., Lindeberg, T., Eckstein, W., Steger, C. & Baumgartner, A. (2000). Automatic extraction of roads from aerial images based on scale space and snakes, *Machine Vision and Application* 12(1): 23–31.

Lin, H.-C., Wang, L.-L. & Yang, S.-N. (1996). Automatic determination of the spread parameter in gaussian smoothing, *Pattern Recognition Letters* 17(12): 1247–1252.

Liu, Z., Cai, J. & Buse, R. (2003). *Handwriting recognition: soft computing and probabilistic approaches*, Springer Verlag, Berlin, Germany.

Manjunath, B. S. & Ma, W. Y. (1998). Texture features for browing and retrieval of image data, *IEEE Transactions on Pattern Analysis and Machine Intelligence* 18(8): 837–842.

Mayer, H. & Steger, C. (1998). Scale-space events and their link to abstraction for road extraction, *ISPRS Journal of Photogrammetry and Remote Sensing* 53(2): 62–75.

Mena, J. B. (2003). State of the art on automatic road extraction for gis update: a novel classification, *Pattern Recognition Letters* 24(16): Pages 3037–3058.

Mohammadzadeh, A., Tavakoli, A. & Zoej, M. J. V. (2006). Road extraction based on fuzzy logic and mathematical morphology from pan-sharpened ikonos images, *Photogrammetric Record* 21(113): 44–60.

Mokhtarzade, M. & Zoej, M. (2007). Road detection from high-resolution satellite images using artificial neural networks, *International Journal of Applied Earth Observation and Geoinformation* 9(1): 32–40.

Musso, A. & Vuchic, V. R. (1988). Characteristics of metro networks and methodology for their evaluation, *Transportation Research Record* (1162): 22–33.

Peng, J. & Jin, Y. Q. (2007). An unbiased algorithm for detection of curvilinear structures in urban remote sensing images, *International Journal of Remote Sensing* 28(23): 5377–5395.

Peng, T., Jermyn, I., Prinet, V. & Zerubia, J. (2008). Incorporating generic and specific prior knowledge in a multi-scale phase field model for road extraction from vhr

Road Feature Extraction from High Resolution Aerial Images Upon Rural Regions Based on Multi-Resolution
Image Analysis and Gabor Filters

195

image, *IEEE Journal of Selected Topics in Applied Earth Observations and Remote Sensing* 1(2): 139–146.

Poullis, C. & You, S. (2010). Delineation and geometric modeling of road networks, *ISPRS Journal of Photogrammetry and Remote Sensing* 65(2): 165–181.

Péteri, R., Couloigner, I. & Ranchin, T. (2004). Quantitatively assessing roads extracted from high-resolution imagery, *Photogrammetric Engineering & Remote Sensing* 70(12): 1449–1456.

Quackenbush, L. J. (2004). A review of techniques for extracting linear features from imagery, *Photogrammetric Engineering & Remote Sensing* 70(12): 1383–1392.

Rellier, G., Descombes, X. & Zerubia, J. (2002). Local registration and deformation of a road cartographic database on a spot satellite image, *Pattern Recognition* 35(10): 2213–2221.

Shi, W. & Zhu, C. (2002). The line segment match method for extracting road network from high-resolution satellite images, *IEEE Transactions on Geoscience and Remote Sensing* 40(2): 511–514.

Shukla, V., Chandrakanth, R. & Ramachandran, R. (2002). Semi-automatic road extraction algorithm for high resolution images using path following approach, *the Indian Conference on Computer Vision, Graphics and Image Processing*, Ahmedabad, India.

Soheilian, B. (2008). *Roadmark reconstruction from stereo-images acquired by a ground-based mobile mapping system*, Ph.d thesis, Université Paris Est.

Steger, C. (1996). *Extracting curvilinear structures: A differential geometric approach*, Vol. 1064, Springer Verlag, pp. 630–641.

Steger, C. (1998). An unbiased detector of curvilinear structures, *IEEE Transactions on Pattern Analysis and Machine Intelligence* 20(2): 113–125.

Steger, C., Glock, C., Eckstein, W., Mayer, H. & Radig, B. (1995). *Model-based road extraction from images*, Birkhäuser Basel, Birkhauser Verlag, Basel, Switzerland, pp. 275–284.

Steger, C., Mayer, H. & Radig, B. (1997). *The role of grouping for road extraction*, Vol. 245-256, Birkhäuser, Basel, Switzerland, pp. 1931–1952.

Stoica, R., Descombes, X. & Zerubia, J. (2004). A gibbs point process for road extraction from remotely sensed images, *International Journal of Computer Vision* 57(2): 121–136.

Tournaire, O. & Paparoditis, N. (2009). A geometric stochastic approach based on marked point processes for road mark detection from high resolution aerial images, *ISPRS Journal of Photogrammetry and Remote Sensing* 64(6): 621–631.

Trinder, J. C. & Wang, Y. (1998). Automatic road extraction from aerial images, *Digital Signal Processing* 8: 125–224.

Tupin, F., Maitre, H., Mangin, J.-F., Nicolas, J.-M. & Pechersky, E. (1998). Detection of linear features in sar images: application to road network extraction, *IEEE Transactions on Geoscience and Remote Sensing* 36(2): 434–453.

Valero, S., Chanussot, J., Benediktsson, J., Talbot, H. & Waske, B. (2010). Advanced directional mathematical morphology for the detection of the road network in very high resolution remote sensing images, *Pattern Recognition Letters* 31(10): 1120–1127.

Vapnik, V. N. (1998). *Statistical learning theory*, John Wiley & Sons, Inc., New York.

Wang, F. & Newkirk, R. (1988). A knowledge-based system for highway network extraction, *IEEE Transactions on Geoscience and Remote Sensing* 26(5): 525 – 531.

Wang, P. & Zhang, Y. (1989). A fast and flexible thinning algorithm, *IEEE Transactions on Computers* 38(5): 741–745.

Wang, R. (2004). *Automated road extraction from high-resolution satellite imagery*, Master thesis, University of New Brunswich, Fredericton, Canada.

Wiedemann, C., Heipke, C., Mayer, H. & Olivier, J. (1998). *Empirical evaluation of automatically extracted road axes*, IEEE Computer Society Press, Silver Spring, MD, pp. 172–187.

Wiedemann, C. & Mayer, H. (1996). Automatic verification of roads in digital images using profiles, *Mustererkennung* pp. 609 – 618.

Wilson, H., Mcglone, J. C., MaKeown, D. M. & Irvine, J. M. (2004). User-centric evaluation of semi-automated road network extraction, *Photogrammetric Engineering & Remote Sensing* 70(12): 1353–1364.

Ye, F., Lin, S. & Tang, j. (2006). Automatic road extraction using partical filters from high resolution images, *Journal of China University of Mining and Technology* 16(4): 490–493.

Zhang, C. (2003). *Updating of cartographic road databases by image analysis*, PhD thesis, Swiss Federal Institute of Technology, Zurich, Switzerland.

Zhang, C. (2004). Towards an operational system for automated updating of road databases by integration of imagery and geodata, *ISPRS Journal of Photogrammetry and Remote Sensing* 58(3-4): 166–186.

Zhang, Q. & Couloigner, I. (2004). Automatic road change detection and gis updating from high spatial remotely-sensed imagery, *Geo-Spatial Information Science* 7(2): 89–95.

Zhao, H., Kumagai, J., Nakagawa, M. & Shibasaki, R. (2002). Semi-automatic road extraction from high-resolution satellite image, *Proceedings of Photogrammetric Computer Vision*, Graz, Austria.

Zhou, J., Bischofa, W. F. & Caelli, T. (2006). Road tracking in aerial images based on human-computer interaction and bayesian filtering, *ISPRS Journal of Photogrammetry and Remote Sensing* 61(2): 108–124.

Zhou, J., Cheng, L. & Bischof, W. F. (2007). Online learning with novelty detection in human-guided road tracking, *IEEE Transactions on Geoscience and Remote Sensing* 45(12): 3967–3977.

Zhu, C., Shi, W., Peraresi, M., Liu, L., Chen, X. & King, B. (2005). The recognition of road network from high-resolution satellite remotely sensed data using image morphological characteristics, *International Journal of Remote Sensing* 26(24): 5493–5508.

9

Progress Research on Wireless Communication Systems for Underground Mine Sensors

Larbi Talbi[1], Ismail Ben Mabrouk[1] and Mourad Nedil[2]
[1]Université du Québec en Outaouais
[2]Université du Québec en Abitibi-Témiscamingue
Canada

1. Introduction

After a recent series of unfortunate underground mining disasters, the vital importance of communications for underground mining is underlined one more time. Establishing reliable communication is a very difficult task for underground mining due to the extreme environmental conditions. Nevertheless, wireless sensors are considered to be promising candidates for communication devices for underground mine environment. Hence, they can be useful for several applications dealing with the mining industry such as Miners' tracking, prevention of fatal accident between men and vehicles, providing warning signals when miner entering the unsafe area, monitoring underground gases, message communication, etc.

Despite its potential advantages, the realization of wireless sensors is challenging and several open research problems exist. In fact, underground communication is one of the few fields where the environment has a significant and direct impact on the communication performance. Furthermore, underground mines are very dynamic environments. As mines expand, the area to be covered expands automatically.

In mine, communication requires complete coverage inside the mine galleries, increasing system reliability and higher transmission rates for faster data throughput. It is extremely important for information to be conveyed to and gathered from every point of mine due to both safety and productivity reasons. In order to meet these needs, the communications industry has looked to Ultra-Wide-band (UWB) for wireless sensors. There have been numerous research results in the literature to indicate that UWB is one of the enabling technologies for sensor network applications [1, 2, 3, 4, 5, 6]. Therefore, UWB provides a good combination of high performance with low complexity for WSN applications [7, 8, 9, 10].

Since UWB has excellent spatial resolution it can be advantageously applied in the field of localization and tracking [11, 12, 13]. In addition to UWB technology, multiple antenna systems have drawn great interest in the wireless community. Multiple antenna systems employ multiple antennas at the transmitter, receiver, or both. By using the antennas in a smart fashion, it may be possible to achieve array gain or diversity gain when multiple antennas are located at either the transmitter or receiver link ends. When multiple antennas are present at both link ends, however, the achievable data rate can potentially be increased linearly proportional to the minimum of the number of antennas at the link ends.

In a sensor network, nodes are generally densely deployed. They do not compete with each other but collaborate to perform a common task. Consider a situation where multiple nodes sense the same object and feed the measurements to a remote data fusion center (relay station). Since nodes are spatially clustered, it is natural to let them cooperate as multiple inputs in transmission and receiving, for the ultimate objective to save energy. In [14], Cui, Goldsmith and Bahai investigated the energy efficiency of MIMO and cooperative MIMO techniques in sensor networks. They mainly consider using MIMO for diversity gain, which improves the quality of the link path.

This chapter will study the application of UWB and MIMO techniques in wireless sensor networks. Hence, a channel characterization of the wireless underground channel is essential for the proliferation of communication protocols for wireless sensor network.

2. UWB channel characterization

2.1 Description of the underground mining environment

The measurements were performed in various galleries of a former gold mine, at a 70 m underground level. The environment mainly consists of very rough walls and the floor is not flat and it contains some puddles of water. The dimension of the mine corridors varies between 2.5 m and 3 m in width and approximately 3 m in high. The measurements were taken in both line of sight (LOS) and non line of sight (NLOS) scenarios. Figure 1 illustrates photography of the underground gallery and the measurement arrangement.

Fig. 1. Photography of the Underground Gallery and the Measurement Arrangement.

2.2 Measurement campaign

The transmitter antenna was always located in a fixed position, while the receiver antenna was moved throughout along the gallery on 49 grid points. As shown in figure 2, the grid was arranged as 7X7 points with 5 cm spacing between each adjacent point. The 5 centimetres corresponds to half of wavelength of the lowest frequency component for uncorrelated small scale fading. During all measurements, the heights of the transmitting and receiving antennas were maintained at 1.7 m in the same horizontal level, and the channel was kept stationary by ensuring there was no movement in the surrounding environment.

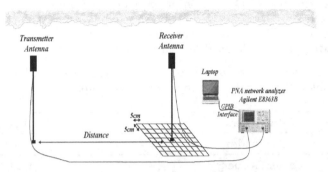

Fig. 2. Overview of the Measurement Setup

The UWB measurements were performed in frequency domain using the frequency channel sounding technique based on S21 parameter obtained with a network analyzer. In fact, the system measurement setup consists of E8363B network analyzer (PNA) and two different kinds of antennas, with directional and omnidirectional radiation patterns, respectively. There were no amplifiers used during the measurements because the distance between the transmitter and the receiver was just 10 meters. The transmitting port of the PNA swept 7000 discrete frequencies ranging from 3 GHz to 10 GHz uniformly distributed over the bandwidth, and the receiving port measured the magnitude and the phase of each frequency component. Figure 3 shows a typical complex channel transfer function (CTF) measured with the Network Analyzer.

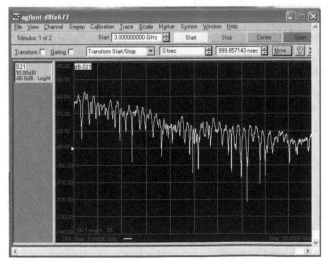

Fig. 3. Channel Transfer Function Measured with the Agilent E8363B Network Analyzer

The frequency span of 1 MHz is chosen small enough so that diffraction coefficients, dielectric constants, etc., can be considered constant within the bandwidth of 7 GHz [15]. At each distance between the transmitter and the receiver, the channel transfer function was measured 30 times, to reduce the effects of random noise on the measurements, and then stored in a computer hard drive via a GPIB interface. The 7 GHz bandwidth gives a theoretical time resolution of 142.9 ps (in practice, due to the use of windowing the time resolution is estimated to be 2/bandwidth) and the sweeping time of the network analyzer is decreased to validate the quasi- static assumption of the channel. The frequency resolution of 1 MHz gives maximum delay range of 1 µs.

Before the measurements, the calibration of the setup was done to reduce the influence of unwanted RF cables effects. Table 1 lists the parameters setup.

Parameters	Values
Bandwidth	7 GHz
Center Frequency	6.5 GHz
Frequency Sweeping Points	7000
Frequency Resolution	1 MHz
Time Resolution	286 ps
Maximum Delay Range	1000 ns
Sweep Average	30
Tx-Rx Antennas Height	1.7 m

Table 1. Measurement System Parameters

Since the measurements are performed in frequency domain, the inverse Fourier transform (IFT) was applied to the measured complex transfer function using Kaiser-Bessel window in order to obtain the channel impulse response. The Kaiser window is designed as FIR filter with parameter β=6 to reduce the side lobes of the transformation.

2.3 Measurements results and analysis

The large scale measurements are performed to determine the propagation distance-power law in the underground environment. The average path loss in dB for arbitrary transmitter-receiver separation distance d can be represented as:

$$PL_{average}(d) = \frac{1}{M}\frac{1}{N}\sum_{i=1}^{M}\sum_{j=1}^{N}\left|H\left(f_i,d\right)\right|^2 \tag{1}$$

where H(f_i,d) is the measured complex frequency response and N represents the number of data points measured during a sweep of 7000 discrete frequencies ranging from 3 GHz to 10 GHz, and M represents the number of sweeps that has been averaged.

According to the measured channel transfer function and the data fitting using the linear least squares regression, the computations of different transmitter-receiver antennas combination have shown that the path loss PL (d) in dB at any location in the gallery can be written as a random log-normal distribution by :

$$PL_{dB}(d) = PL_{dB}(d_0) + 10.n.\log_{10}\left(\frac{d}{d_0}\right) + X_\sigma \qquad (2)$$

where $PL(d_0)$ is the path loss at the reference distance d_0 set to 1m, n is the path loss exponent and X_σ is a zero-mean Gaussian distributed random variable in dB with the standard deviation.

2.3.1 LOS scenario

2.3.1.1 Path loss model

The measurements of UWB propagations channel in line of sight case were made between 1 m and 10 m with intervals of 1 m. Figure 4 illustrates the gallery layout and the measurements Tx-Rx arrangements under LOS and Figure 5 shows the results of path loss as function of distance for the three antennas combinations: directional - directional, directional-omni and omni-omni.

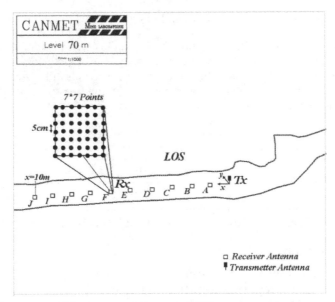

Fig. 4. Gallery Layout and Measurement Setup in LOS

As listed in Table 2, the path loss exponent n, in LOS scenario is equal to 1.99, 2.01 and 2.11 for directional-omni, directional-directional, and omni-omni antennas combination respectively. It can be noted that the path loss exponent for all these combinations is close to free space path loss exponent where n=2, with the smallest path loss fluctuation for directional-omni antenna combination, and the standard deviation of Gaussian random variable σ_{dB} is smaller for directional antenna in LOS environment. The results of path loss exponent values observed in [16] [17] for indoor UWB propagation are lower to the results

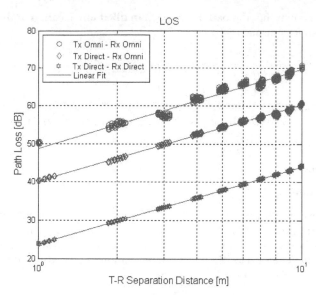

Fig. 5. Path Loss vs. T-R Separation Distance in LOS

obtained for underground UWB propagation. In an indoor environment, such as a corridor or a hallway clear of obstacles, the results may show lower path loss exponent due to multipath signal addition, whereas in the mine gallery, the walls are uneven, scattering the signal and thus showing results in closer agreement with the free-space path loss exponent, due mainly to the LOS component reaching the antenna.

LOS	Omni – Omni	Direct – Direct	Direct – Omni
n	2.11	2.01	1.99
σ_{dB}	0.89	0.13	0.32

Table 2. Summary of Path Loss Exponents n and Standards Deviations σ_{dB} in LOS.

2.3.1.2 RMS delay spread

A statistical characterization of the channel impulse response is a useful process for describing the rapid fluctuations of the amplitude, phase, and multipath propagation delays of the UWB signal. The number of multipath in an underground environment is more important due to the reflection and scattering from the ground and surrounding rough surfaces. Figure 6 shows a typical power delay profile (PDP) measured with omni-omni antenna in LOS environment.

In order to compare different multipath channels of different antennas combination, the mean excess delay and RMS delay spread are evaluated using the below equations [18] :

-   RMS delay spread is the square root of the second central moment of the power delay profile given by:

$$\tau_{rms} = \sqrt{\overline{\tau^2} - \left(\overline{\tau}\right)^2} \tag{3}$$

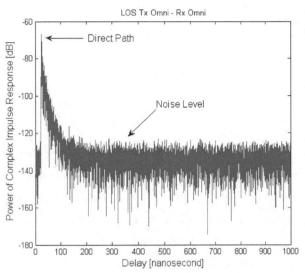

Fig. 6. Typical underground Power Delay Profile in LOS

- Mean excess delay is the first moment of the power delay profile defined by:

$$\overline{\tau} = \frac{\sum_k a_k^2 . \tau_k}{\sum_k a_k^2} = \frac{\sum_k P(\tau_k) . \tau_k}{\sum_k P(\tau_k)} \tag{4}$$

$$\overline{\tau^2} = \frac{\sum_k a_k^2 . \tau_k^2}{\sum_k a_k^2} = \frac{\sum_k P(\tau_k) . \tau_k^2}{\sum_k P(\tau_k)} \tag{5}$$

Where a_k, $P(\tau_k)$ and τ_k are the gain, power and delay of the kth path respectively. From (3), (4) and (5) we have calculated the RMS delay spread for each antenna combination by using predefined thresholds. A threshold of 40 dB below the strongest path was chosen to avoid the effect of noise on the statistics of multipath arrival times. Fig. 7 shows the effects of antenna directivity on the RMS delay spread computed from the cumulative distribution function in LOS scenario.

According to the figure 7, we can observe that for 50% of all locations, the directional - directional combination offers the best result of τ_{rms} with 2 ns. However, the directional-omni and the omni-omni combinations introduce 7.7 ns and 9.5 ns of τ_{rms} respectively. Hence, we can say that the former combination reduces 7.5 ns of τ_{rms} in comparison with the latter one. The effect of directional antenna in underground LOS environment is similar to the results reported in indoor channel [19] [20].

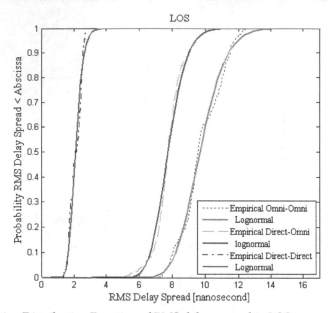

Fig. 7. Cumulative Distribution Function of RMS delay spread in LOS

2.3.2 NLOS scenario

2.3.2.1 Path loss model

The measurements of UWB propagations in non line of sight were made between 4 m and 10 m with intervals of 1m. Figure 8 illustrates the gallery layout and the measurements arrangement in NLOS.

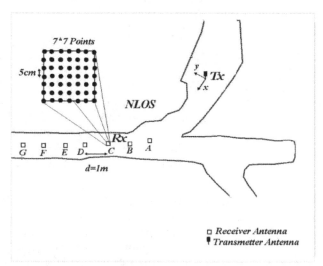

Fig. 8. Gallery Layout and Measurement Setup in NLOS

The results of path loss as function of distance for directional - directional and omni - omni antennas combinations are shown in Figure 9.

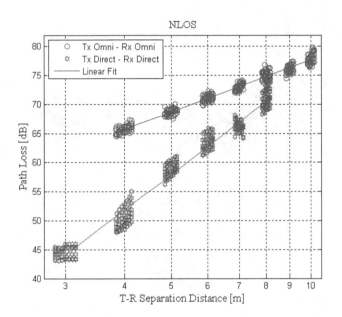

Fig. 9. Path Loss vs. T-R Separation Distance in NLOS

As listed in Table 3, the path loss exponent with directional antennas is twice larger than of the omnidirectional antennas.

NLOS	Omni – Omni	Direct – Direct
n	3.00	6.16
σ_{dB}	0.66	1.47

Table 3. Summary of Path Loss Exponents n and Standards Deviations σ_{dB} in NLOS

2.3.2.2 RMS delay spread

In NLOS scenario, the UWB signal reaches the receiver through reflections, scattering, and diffractions. Figure 10 shows that a typical power delay profile (PDP) measured with Omni-Omni antenna in NLOS environment consists of components from multiple reflected, scattered, and diffracted propagation paths.

Figure 11 shows that the use of directional antennas, for 50% of all locations in NLOS scenario, can reduce, 13 ns of τ_{rms} compared to omnidirectional antennas.

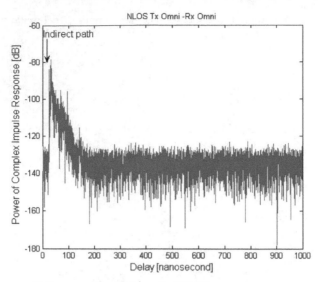

Fig. 10. Path Loss vs. T-R separation distance in NLOS

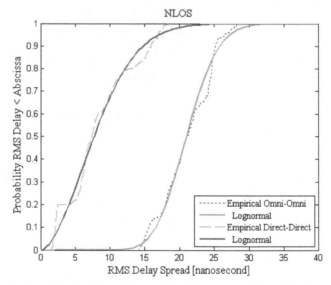

Fig. 11. Cumulative Distribution Function of RMS delay spread in NLOS

3. MIMO channel characterization at 2.4 GHz

3.1 Description of the underground environment

Measurements were conducted in a gallery located at a 40-m deep underground level.

In this gallery, the floor is uneven with bumps and several ditches. In addition, the walls are not aligned. Dimensions vary almost randomly throughout the gallery, although the latter is supposed to have a width of about 4 to 5 m. The gallery also has several branches of different size at variant locations. The humidity is still high, drops of water falling from everywhere and big pools of water cover the ground. The temperature is stable of 6 to 15° C along the year. A photography of this underground gallery is shown in figure 12.

Fig. 12. Photography of the mine gallery

3.2 Measurement setup

The MIMO antenna system consists of a set patch antenna, developed in our laboratory and have been used for transmission and reception of the RF signal, at 2.4GHz. Measurement campaigns under LOS and NLOS scenarios were performed in frequency domain using the frequency channel sounding technique based on measuring S_{21} parameter with a network analyzer (Agilent E8363B). In fact, the system measurement setup, as shown in figure 13, consists of a network analyzer (PNA), 2X2 MIMO antenna set, two switches, one power amplifier for the transmitting signal and one low noise amplifier for the receiving signal. Both amplifiers have a gain of 30 dB.

For the Line-of-Sight (LOS) scenario, the transmitter remained fixed at T_{x1} , where the receiver changed its position along the gallery, from 1 meter up to 25 meters far from the transmitter. While for NLOS the transmitter remained fixed at T_{x2} and the $T_x - R_x$ separation varies from 6m up to 25m. Figure 14 illustrates photography of the receiver location and a map of the underground gallery.

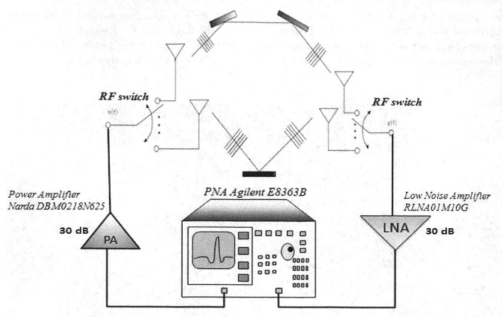

Fig. 13. Measurement setup

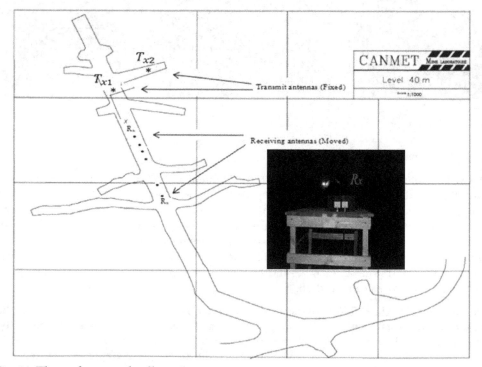

Fig. 14. The underground gallery plan

3.3 Measurement results

3.3.1 RMS delay (τ_{RMS})

The RMS delay spread roughly characterizes the multipath propagation in the delay domain. The RMS delay spread is the square root of the second central moment of the averaged power and it is defined as:

$$\tau_{rms} = \sqrt{\overline{\tau^2} - (\overline{\tau})^2} = \sqrt{\frac{\sum_i P_i \tau_i^2}{\sum_i P_i} - \left(\frac{\sum_i P_i \tau_i}{\sum_i P_i}\right)^2} \tag{6}$$

where $\overline{\tau}$ is the mean excess delay, $\overline{\tau^2}$ is the average power and P_i is the received power (in linear units) at τ_i corresponding arrival time. We have a threshold of 10 dB for all power delay profiles, in order to guarantee the elimination of the noise.

The RMS delay spread has been computed for each impulse response of all the gallery measurements using the 2X2 MIMO system under LOS and NLOS scenarios and plotted in terms of the separation distance d_{Tx-Rx} in figure 15.

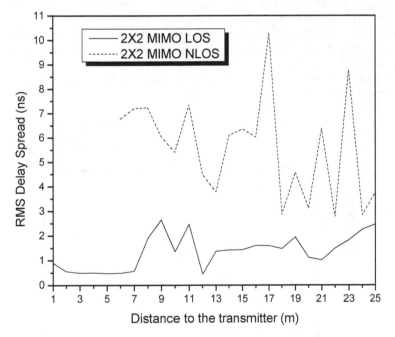

Fig. 15. RMS delay spread as a function of the distance

For the considered underground gallery, the profile seen in figure 15 is not monotonically increasing as may be expected. Results thus show propagation behavior that is specific for these underground environments. This is likely due to scattering on the rough sidewalls' surface that exhibit a difference of 25 cm between the maximum and minimum surface variation. Moreover, the RMS delay for the MIMO in NLOS scenario is higher than the one of MIMO by about 5 ns due to the walls attenuation. Table 4 summarizes the RMS values for LOS and NLOS locations.

RMS (ns)	MIMO LOS	MIMO NLOS
Minimum / Maximum	0.44 / 2.64	2.7815 / 10.292
Mean / Standard deviation (σ)	1.33 / 0.68	5.6081 / 2.0750

Table 4. Summary of the RMS delay spread for measurements corresponding to LOS and NLOS galleries

3.3.2 Path loss

Path loss in the channel is normally distributed in decibel (dB) with a linearly increasing mean and is modeled as:

$$PL_{dB}(d_0) = \overline{PL_{dB}}\,(d_0) + 10\alpha\log\left(\frac{d}{d_0}\right) + X \tag{7}$$

where $\overline{PL_{dB}}\,(d_0)$ is the mean path loss at the reference distance d_0, $10\alpha\log\,(d/d_0)$ is the mean path loss referenced to d_0, and X is a zero mean Gaussian random variable expressed in dB. Path Loss as a function of distance are shown in figure 16 and figure 17 for both LOS and NLOS galleries respectively. The mean path loss at d_0 and the path loss exponent α were determined through least square regression analysis [21]. The difference between this fit and the measured data is represented by the Gaussian random variable X. Talble 5 lists the values obtained for α and σ_X (standard deviation of X).

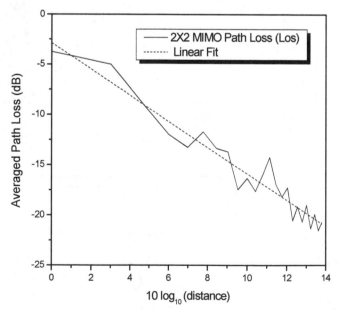

Fig. 16. Average Path versus distance in LOS scenario

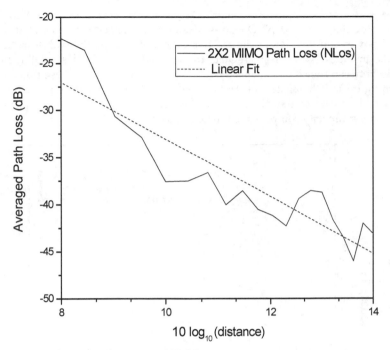

Fig. 17. Average Path versus distance in NLOS scenario

	MIMO LOS	MIMO NLOS
α	1.73	3.03
σ_X	1.29	2.75

Table 5. Path Loss exponent α and standard deviation of X (σ_X)

From the results shown in Table 5 the NLOS scenario have path loss exponent greater than 2 and also have larger σ_X value compared with LOS scenarios. While, for the LOS case the exponent α=1.73, is smaller than the free space exponent α=2, the reason behind this is because of the collection of all multipath components so that a higher power is received than the direct two signals in the free space.

3.3.3 Capacity

If we consider a system composed on m transmitting antennas and n receiving antennas, the maximum capacity of a memoryless MIMO narrow band channel expressed in bits/s/Hz, with a uniform power allocation constraint and in the presence of additional white Gaussian noise is given by Foschini et al.[22]:

$$C = \log_2 \det (I_m + \sigma.HH^H) \tag{8}$$

where σ is the average signal to noise ratio per receiving antenna; I_m denotes the identity matrix of size m, the upper script ^{H} represents the hermitian conjugate of the matrix and det(X) means the determinant of a matrix X. To clearly point out the MIMO system performance for the LOS and NLOS cases, the ergodic capacity is calculated for a fixed transmitted power and the SNR at the receiver is determined by the path loss. In this case, the capacity includes both effects related to received power and spatial richness. The relationship between the channel capacity C and the distance d_{Tx-Rx} based on equation (8) is shown in figure 18. Obviously, one can see that the NLOS suffer from its higher path-loss exponent which is due to the directional radiation pattern of the MIMO patch antenna resulting in lower capacity compared to the LOS case by about 3 bit/s/Hz.

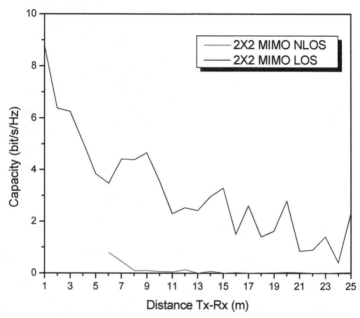

Fig. 18. Channel capacity for LOS and NLOS scenarios

4. Conclusion

This study deals with several aspects relative to UWB and MIMO propagation channel and its deployment for wireless sensors. Successful design and deployment of these techniques require detailed channel characterization. Measurement campaigns, made at two different deep levels in a former gold mine under LOS and NLOS scenarios, have been analyzed to obtain the relevant statistical parameters of the channel.

Although MIMO system can offer high capacity performance through multipath propagation channel but it has some drawbacks such as complexity, power consumption and size limitation of the wireless sensor. However, UWB has several advantages compared to narrowband systems. The wide bandwidth (typically 500 MHz or more) gives UWB excellent immunity to interference from narrowband systems and from multipath effects. Another important benefit of UWB is its high data rate. Additionally, UWB offers significant advantages with respect to robustness, energy consumption and location accuracy.

Nevertheless, UWB technology for wireless networks is not all about advantages. Some of the main difficulties of UWB communication are low transmission power so information can only travel for short distance comparing to 2.4 GHz which can rich long distance. Moreover UWB in the microwave range does not offer a high resistance to shadowing, but this problem can be mitigated in sensor networks by appropriate routing, and possible collaborative communications.

5. References

A. A. M. Saleh and R. A. Valenzuela, "A Statistical Model for Indoor Multipath Propagation," IEEE J. Select. Areas Commun., vol. SAC-5, pp. 128-137, Feb. 1987.

A. F. Molisch, B. Kannan, C. C. Chong, S. Emami, A. Karedal, J. Kunisch, H. Schantz, U. Schuster and K. Siwiak, "IEEE 802.15.4a Channel Model – Final Report", IEEE 802.15-04 0662-00-004a, San Antonio, TX, USA, Nov. 2004.

A.J. Goldsmith S. Cui and A. Bahai.: 'Energy-efficiency of mimo and cooperative mimo in sensor networks', IEEE Journal on Selected Areas of Communications, 22(6), August 2004.

A.Muqaibel, A. Safaai-Jazi, A, Attiya, B Woerner, and S. Riad, "Path-Loss and time dispersion parameters for indoor UWB propagation ", Wireless Communications, IEEE Transactions, Vol 5, Issue 3, March 2006 Pages 550-559.

Arslan A, Chen AN and Benedetto MG (2006) Ultra-wideband wireless communication. Wiley Interscience, Hoboken, New Jersey.

Arslan H and Benedetto MGD (2005) Introduction to UWB. Book Chapter, Ultra Wideband Wireless Communications (ed. Arslan H), John Wiley & Sons, USA.

Chehri A and Fortier P (2006a) Frequency domain analysis of UWB channel propagation in underground mines. Proceedings of IEEE 64th Vehicular Technology Conference, Montreal, Canada, 25–28 September 2006, pp. 1–5.

Chehri A, Fortier P and Tardif PM (2006a) Deployment of ad-hoc sensor networks in underground mines. Proceedings of Conference on Wireless and Optical Communication, and Wireless Sensor Network, Alberta, Canada, 3–4 July 2006, pp. 13–19.

Choi JD and Stark WE (2002) Performance of ultra-wideband communications with suboptimal receivers in multipath channels. IEEE Journal on Selected Areas in Communications, pp. 1754–1766.

F. Granelli, H. Zhang, X. Zhou, S. Maranò, "Research Advances in Cognitive Ultra Wide Band Radio and Their Application to Sensor Networks," Mobile Networks and Applications, Vol. 11, pp. 487-499, 2006.

G. J. Foschini and J. Gans, "On Limits of Wireless Communications in a Fading Environment when Using Multiple Antennas", Wireless Personal Communications, vol. 6, no. 3, pp. 315-335, March, 1996

J. Li, T. Talty, "Channel Characterization for Ultra-Wideband Intra-Vehicle Sensor Networks," Military Communications conference (MILCOM), pp. 1-5, 2006.

L. Stoica, A. Rabbachin, H.O. Repo, T.S. Tiuraniemi, I. Oppermann, "An Ultrawideband System Architecture for Tag Based Wireless Sensor Networks," IEEE Transactions on Vehicular Technology, Vol. 54, pp. 1632-1645, 2005.

L. Yuheng, L. Chao, Y. He, J. Wu, Z. Xiong, "A Perimeter Intrusion Detection System Using Dual-Mode Wireless Sensor Networks," Second International Conference on Communications and Networking in China, pp. 861-865, 2007.

M. Chamchoy, W. Doungdeun, S, Promwong "Measurement and modeling of UWB path loss for single-band and multi-band propagation channel", Communications and Information Technology, 2005. ISCIT 2005. IEEE International Symposium, vol2, 12-14 Oct. 2005 Pages:991-995.

Molisch AF (2005) Ultra wideband propagation channels-theory, measurement, and modeling. IEEE Transactions on Vehicular Technology, pp. 1528–1545.

Molisch, A. F.; Cassioli, D.; Chong, C.-C.; Emami, S.; Fort, A.; Kannan, B.; Karedal, J.; Kunisch, J.; Schantz, H. G.; Siwiak, K.; Win, M. Z.; "A Comprehensive Standardized Model for Ultrawideband Propagation Channels", Antennas and Propagation, IEEE Transactions on. Volume 54, Issue 11, Part 1, Nov. 2006 Page(s):3151 - 3166

Nedil M, Denidni TA, Djaiz A and Habib AM (2008) A new ultra-wideband beamforming for wireless communications in underground mines. Progress in Electromagnetics Research, 4: 1–21.

R.S. Thoma, O. Hirsch, J. Sachs, Zetik, R., "UWB Sensor Networks for Position Location and Imaging of Objects and Environments," The Second European Conference on Antennas and Propagation (EuCAP), pp. 1-9, 2007.

S. Ghassemzadeh, L. Greenstein, T. Sveinsson, A.Kavcic, V. Tarokh, "UWB indoor path loss model for residential and commercial environments," in Proc. IEEE Veh. Technol. Conf (VTC 2003- Fall), Orlando, FL, USA, pp. 629-633, Sept. 2003.

T. S. Rappaport, Wireless Communications: Principles & Practice, Upper Saddle River, NJ, Prentice Hall PTR, 1996

X. Huang, E. Dutkiewicz, R. Gandia, D. Lowe, "Ultra-Wideband Technology for Video Surveillance Sensor Networks," IEEE International Conference on Industrial Informatics, pp. 1012-1017, 2006.

10

Cold Gas Propulsion System – An Ideal Choice for Remote Sensing Small Satellites

Assad Anis
NED University of Engineering and Technology
Pakistan

1. Introduction

Cold gas propulsion systems play an ideal role while considering small satellites for a wide range of earth orbit and even interplanetary missions. These systems have been used quite frequently in small satellites since 1960's. It has proven to be the most suitable and successful low thrust space propulsion for LEO maneuvers, due to its low complexity, efficient use of propellant which presents no contamination and thermal emission besides its low cost and power consumed. The major benefits obtained from this system are low budget, mass, and volume. The system mainly consists of a propellant tank, solenoid valves, thrusters, tubing and fittings (fig. 1). The propellant tank stores the fuel required for attitude control of satellite during its operation in an orbit. The fuel used in cold gas systems is compressed gas. Thrusters provide sufficient amount of force to provide stabilization in pitch, yaw and roll movement of satellite. From design point of view, three important components of cold gas propulsion systems play an important role i.e. mission design, propellant tank and cold gas thrusters. These components are discussed in detail in section 3. Selection of suitable propellant for cold gas systems is as important as above three components. This part is discussed in section 2 of this chapter. Section 4 describes the case study of cold gas propulsion system which is practically implemented in Pakistan's first prototype remote sensing satellite PRSS.

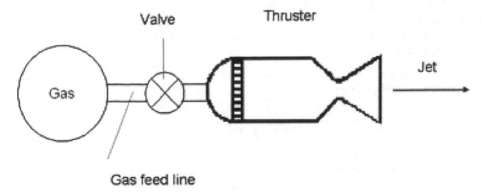

Fig. 1. Schematic of cold gas propulsion system

2. Cold gas propellants

Table 1 shows typical performance values for selected cold gas propellants. Nitrogen is most commonly used as a cold gas propellant, and it is preferred for its storage density, performance, and lack of contamination concerns. As shown in table below, hydrogen and helium have greater specific impulse as compared to other propellants, but have a low molecular weight. This quality causes an increased tank volume and weight, and ultimately causing an increase in system weight. Carbon dioxide can be a good choice, but due to its toxic nature, it is not considered for cold gas systems.

Another good alternative propellant could be ammonia, which stores in its liquid form to reduce tank volume. Its specific impulse is higher than nitrogen or other propellants and reduces concerns of leakage, although it also necessitates a lower mass flow rate. Despite the benefits, ammonia is not suitable for this system as one alternative to decrease the system size and weight includes pressurizing the satellite and allowing the entire structure to act as a propellant tank, as previously mentioned. In this system, the ammonia could cause damage to electrical components.

Propellant	Molecular Weight (Kg/Kmole)	Density (g/cm³)	Specific Thrust (s)	
			Theoretical	Measured
Hydrogen	2.0	0.02	296	272
Helium	4.0	0.04	179	165
Nitrogen	28.0	0.28	80	73
Ammonia	17.0	Liquid	105	96
Carbon dioxide	44.0	Liquid	67	61

Table 1. Cold Gas Propellant Performances

3. Cold gas propulsion system design

3.1 Mission design

In order to design a cold gas propulsion system for a specific space mission, it is important first to find out the ΔV requirements for the maneuvers listed in table 2. Table 2 gives information about all the operations performed for spacecraft attitude and orbit control. However, cold gas systems are used only for attitude control and orbit maintenance and maneuvering (table 3).

Tsiolkowski equation and its corollaries are used to convert these velocity change requirements into propellant requirements.

$$\Delta V = g_c I_{sp} \ln\left(\frac{W_i}{W_f}\right) \tag{1}$$

$$W_f = W_i\left[1 - \exp\left(-\frac{\Delta V}{g_c I_{sp}}\right)\right] \tag{2}$$

Task	Description
Mission Design	(Translational velocity change)
Orbit changes	Convert one orbit to another
Plane changes	
Orbit trim	Remove launch vehicle errors
Stationkeeping	Maintain contellation position
Repositioning	Change constellation position
Attitude Control	(Rotational velocity change)
Thrust vector control	Remove vector errors
Attitude control	Maintain an attitude
Attitude changes	Change attitudes
Reaction wheel unloading	Remove stored momentum
Maneuvering	Repositioning the spacecraft axes

Table 2. Spacecraft Propulsion Functions

Propulsion Technology	Orbit Insertion		Orbit Maintenance and Maneuvering	Attitude Control	Typical Steady State I_{sp} (S)
	Perigeee	Apogee			
Cold Gas			Yes	Yes	30-70
Solid	Yes	Yes			280-300
Liquid					
Monopropellant			Yes	Yes	220-240
Bipropellant	Yes	Yes	Yes	Yes	305-310
Dual Mode	Yes	Yes	Yes	Yes	313-322
Hybrid	Yes	Yes	Yes		250-340
Electric		Yes	Yes		300-3,000

Table 3. Principal Options for Sapcecraft Propulsion Systems

$$W_p = W_f \left[\exp\left(\frac{\Delta V}{g_c I_{sp}} \right) - 1 \right] \quad (3)$$

In case of cold gas propulsion systems, the pressure, mass, volume and temperature of the propellant are interconnected by general gas equation.

$$PV = mRT \quad (4)$$

3.2 Tank design

Satellite propellant tanks used in cold gas propulsion systems are either spherical or cylindrical in shape. Tank weights are a byproduct of the structural design of the tanks. The load in the walls of the spherical pressure vessels is pressure times the area as shown in figure 2. The force PA tending to separate the tanks is given as,

$$PA = P\pi r^2 \tag{5}$$

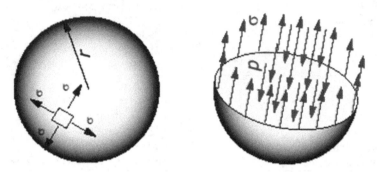

Fig. 2. Spherical Tank Stress

Stress σ is calculated as,

$$stress = \sigma = \frac{load}{area} = \frac{P\pi r^2}{2\pi rt} = \frac{Pr}{2t} \tag{6}$$

The thickness of the tank is accurately calculated by including joint efficiency in eq. (7) and is given as follows,

$$t = \frac{P \times r}{2\sigma e - 0.2P} \tag{7}$$

In case of cylindrical pressure vessel, the hoop stress is twice that in spherical pressure vessels. The longitudinal stresses in cylindrical pressure vessels remain the same as in spherical pressure vessels. To determine the hoop stress σ_h, a cut is made along the longitudinal axis and construct a small slice as illustrated in figure 3.

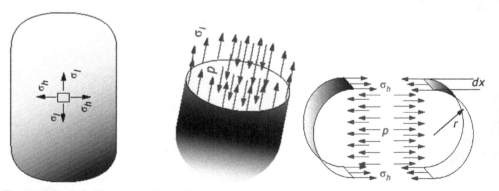

Fig. 3. Cylindrical Pressure Vessel Stresses

The equation may be written as,

$$2.\sigma_h.t.d_x = p.2.r.d_x$$

$$\sigma_h = \frac{pr}{t} \tag{8}$$

3.3 Thrusters design

Thrusters are the convergent-divergent nozzles (fig. 4) that provide desired amount of thrust to perform maneuvers in space. The nozzle is shaped such that high-pressure low-velocity gas enters the nozzle and is compressed as it approaches smallest diameter section, where the gas velocity increases to exactly the speed of sound.

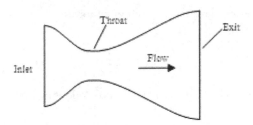

Fig. 4. Convergent-Divergent Nozzle

Thrust is generated by momemtum exchange between the exhaust and the spacecraft and by the pressure imbalance at the nozzle exit. According to Newton's secomd law the thrust is given as

$$F = \dot{m}V_e \tag{9}$$

or we may write as,

$$F = \frac{\dot{w}_p}{g}V_e \tag{10}$$

$$F = P_e A_e \tag{11}$$

In case of satellites, the thrusters are designed for infinite expansion i.e. for vacuum conditions where ambient pressure is taken as zero. The thrust equation for infinite expansion is given as,

$$F = A_t P_c \gamma \left[\left(\frac{2}{\gamma-1} \right)\left(\frac{2}{\gamma+1} \right)\left(1 - \frac{P_e}{P_c} \right) \right] + P_e A_e \tag{12}$$

The area ratio and pressure ratio is given as,

$$\frac{A_e}{A_t} = \frac{1}{M_e}\left\{\left(\frac{2}{\gamma+1}\right)\left(1+\frac{\gamma-1}{2}M_e^2\right)\right\}^{\frac{\gamma+1}{2\gamma-1}}$$ (13)

$$\frac{P_e}{P_c} = \left(1+\frac{\gamma-1}{2}M_e^2\right)^{\frac{\gamma}{\gamma-1}}$$ (14)

The specific impulse (I_{sp}) for cold gases ranges from 30-75 seconds and may be calculated as,

$$I_{sp} = \frac{C^*}{g}\gamma\left\{\left(\frac{2}{\gamma-1}\right)\left(\frac{2}{\gamma+1}\right)^{\frac{\gamma+1}{\gamma-1}}\left(1-\frac{P_e}{P_c}\right)^{\frac{\gamma-1}{\gamma}}\right\}^{\frac{1}{2}}$$ (15)

Pressure at throat can be calculated by the following formula

$$\frac{P_t}{P_c} = \left(1+\frac{\gamma-1}{2}\right)^{-\frac{\gamma}{\gamma-1}}$$ (16)

The characteristics velocity (C^*) can be calculated by following formula

$$C^* = \frac{a_0}{\gamma\left(\frac{2}{\gamma+1}\right)^{\frac{\gamma+1}{2(\gamma-1)}}}$$ (17)

The exite velocity is given as

$$V_e = \sqrt{\frac{2\gamma R T_c}{\gamma-1}\left(1-\frac{P_e}{P_c}\right)^{\frac{\gamma-1}{2}}}$$ (18)

The above equations are helpful in designing of a cold thruster.

4. Case study

The author has personally leaded and guided the satellite Research and Development Centre research team of Pakistan Space and Upper Atmosphere Research Commission in designing and development of cold gas propulsion system of prototype of Pakistan's first remote sensing satellite (PRSS). Satellite research and development center Karachi has produced an inexpensive and modular system for small satellites applications. The cold gas propulsion resulting from the effort is unique in several ways. It utilizes a simple tank storage system in which the entire system operates at an optimum design in line pressure. In order to minimize the power consumption, the thrusters are operated by solenoid valves that require an electric pulse to open and close. Between the pulses the thruster is magnetically latched in either the open or closed position as required. This dramatically reduces the power required by the thruster valves while maintaining the option for small impulse bit. Flow rate sensors are used in the system in order to avoid any failure i.e. complete pressure lost during opened valve position. The system uses eight Thrusters of 1N each functioning with inlet pressure of 8 bars. By integrating these thrusters to the spacecraft body, pitch, yaw and roll control as well as ΔV can be accomplished. The choice of suitable propellant also plays an important role in designing cold gas systems. Compressed nitrogen gas offers a very good combination of storage density and specific impulse, as compared with other available cold gas propellants. The use of Hydrogen or helium requires much larger mass, because of their low gas density. Since the propellant is simple pressurized nitrogen, a variety of suitable tank materials can be selected. The tank designed and developed for this mission is Aluminum 6061 spherical tank which stores 2 kg of gaseous nitrogen. The whole system is well tested before mounting on the honeycomb PRSS structure.

4.1 Introduction to PRSS

PRSS is a prototype satellite which is not developed for flight in future. The purpose of this work is to design, develop and test a small satellite on ground so that the experience can be utilized in near future on engineering qualified and flight models. The CAD model of PRSS is shown in fig. 5. This model is developed in PRO/E wildfire 2.0 software.

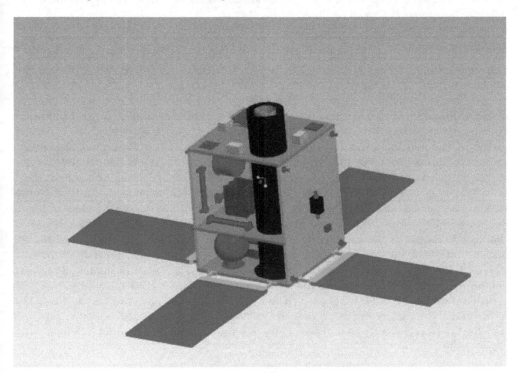

Fig. 5. CAD model of PRSS

The satellite mainly consists of

- 1 Telescope

- CCD Camera
- Optics Electronics
- Cold Gas Propulsion System which includes 8
- thrusters, 1 propellant tank and regulators and
- fittings
- 3 Reaction wheels
- 3 Gyros
- 3 Torque Rods
- 3 Digital Sun Sensor DSS
- RF systems & Antennas
- On Board Computer Electronics
- Power System
- 4 Solar Panels on each side of the cube
- Honeycomb Aluminum Structure

All the above mentioned systems have been integrated successfully on Al 6061 honeycomb structure which is cubical in shape with dimensions 1 m x 1m x 1.2 m. All subsystems have been designed for 3 years satellite life. PRSS has an overall weight of 100 kg and therefore falls into the category of small satellites.

4.2 System design

Cold gas propulsion systems use thrusters which utilize smallest rocket technology available today. These systems are well known for their low complexity when characterized by low specific impulse. They are the cheapest, simplest and reliable propulsion systems available for orbit maintenance and maneuvering and attitude control. Cold gas Propulsion systems are designed for use as satellite maneuvering control system where a limited lifetime is required. Their specific impulse ranges from 30 seconds to 70 seconds, depending on the type propellant used. They usually consist of a pressurized gas tank, control valves, regulators, filters and a nozzle. The nozzle can be of bell type, conical, or a tube nozzle. SRDC-K will be using a standard conical nozzle, with a 16^0 half-angle and nozzle area ratio of 50:1. A schematic of a cold gas thruster system used by PRSS is shown below in Fig 6. System weight is mainly determined by the pressure in the thrust chamber. The increased chamber pressure results in increase propellant tank and piping masses, therefore, an optimum pressure must be used so that the system weight can be minimized. Nitrogen is stored at 100 bar pressure in propellant tank. Fill and drain valves facilitates filling and venting nitrogen from the system. Eight Thrusters are connected to solenoid valves and propellant tank with PTFE tubing which can carry a pressure of more than 20 bars. The inline and the thrusters operating pressure is 8 bars. The system also contains pressure transducer before and after pressure regulator to sense the tank pressure and inline pressure respectively.

4.3 Propellant tank design, development and testing

The propellant tank as shown in fig. 7 is a standard spherical pressure vessel. It is being designed and built by SRDC-K, with the detailed analysis also being performed at SRDC-K. In order to reduce costs the tank is being welded by two hemispherical Aluminum parts.

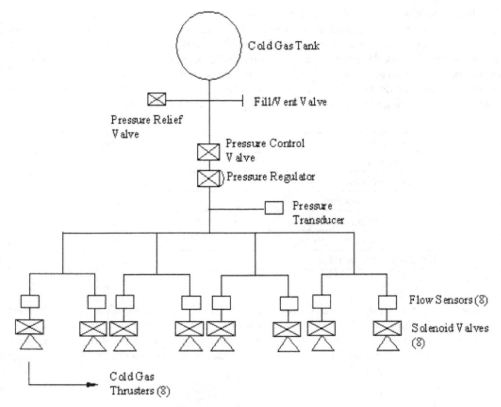

Fig. 6. Cold Gas Propulsion System for PRSS

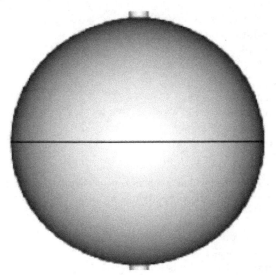

Fig. 7. Propellant Tank

The hemisphere has a wall thickness of 4.2 mm and a factor of safety of 1.5 is used. This gives a minimum theoretical burst pressure of 200 bars. ASME section VIII pressure vessel code is used for the designing the spherical propellant tank. Table 4 presents the calculated values of design tank design parameters. Titanium could have been another choice for the designing of propellant tank. The space grade of titanium is TiAl4V. The weight of the propellant tank would have been less with titanium as compared with aluminum but the cost in manufacturing titanium tank is much higher as compared with aluminum.

Parameters	Designed Parameters
Propellant	N_2 Gas
Tank volume	$0.016 \ m^3$
Operating pressure	100 bars
Proof pressure	150 bars
Burst pressure	200 bars
Thickness of tank shell	0.0042 m

Table 4.

The tank design analyses included stress analysis for the tank shell. This approach used assumptions, computer tools, test data and experimental data which are commonly utilized on a majority of the pressure vessels for successful design, fabrication, testing and qualification. The following factors have been taken in to consideration for performing stress analysis on the tank shell.

- Temperature environment
- Material properties
- Volumetric properties
- Mass properties of the tank shell material
- Mass properties of fluid
- Fluids used by the tank
- External loads
- Size of girth weld
- Resonant frequency
- Tank boundary conditions
- Residual stress in girth weld
- Load reaction points and
- Design safety factors

The validation of tank shell design has been done by stress analysis and also the resonant frequencies have been obtained. The propellant tank is subjected to the following sequence of acceptance tests,

- Preliminary visual examination
- Ambient proof pressure test
- External leakage test
- Penetrant inspection

- Radiographic inspection
- Mass measurement
- Final examination
- Cleanliness verification

The ambient hydrostatic proof pressure test is conducted at 130 +20/-0 bars for a pressure hold period of 300 seconds. Post acceptance test, radiographic inspection of the girth weld and penetrant inspection of the entire external surface are conducted to verify that the tank is not damages during acceptance testing. All units successfully passed acceptance testing. After the conclusion of acceptance testing one propellant tank was subjected to the following sequence of qualification tests prior to delivery:

- Proof pressure cycling test
- MEOP pressure cycling test
- External Leakage test
- Radiographic inspection
- Penetrant inspection
- Burst pressure test
- Visual inspection
- Data review

The propellant tank assembly has successfully completed all acceptance and qualification level testing. The tank meets or exceeds all requirements that provide the low cost solution to the spacecraft.

After successful testing, propellant tank is then mounted on PRSS structure as shown in fig. 8.

Fig. 8. Installation of Propellant Tank on PRSS Structure

4.4 Thrusters design, development and testing

This system uses 8 thrusters (fig 9.a) of 1N mounted on PRSS as shown in fig. 10. These thrusters have been designed and developed for infinite expansion i.e. for vacuum

(a) Cold Gas Thruster   (b) Test Bench

Fig. 9.

Fig. 10. PRSS Structure

conditions and hence, the atmospheric pressure is zero. Area ratio of 50 has been used while the combustion chamber pressure is 8 bars. The characteristics velocity has been calculated and equals to 433.71 m/sec and as result of that the I_{sp} came out to be 73 seconds. Assuming a nozzle efficiency of 98% the nozzle cone half angle has been calculated as 16^0. Thrusters have been developed using stainless steel material. The use of the stainless steel eliminates the potential for reaction between propellant and thruster and also outgassing concerns. The test bench developed at S/P/T laboratory as shown in fig.9.b is capable of testing cold gas thrusters from 1 to 5N. The system consists of an aluminum plate which is mounted on a ball bearing. The thruster is connected to a plate and fitted with solenoid valve through SS 316 tubing.

The system uses FUTEK load cell which is basically a force sensor to measure the force from the thruster. Pressure data logger and transducer are also connected to the system to measure the pressure during testing.

4.5 Propulsion system integration on PRSS structure

All components of propulsion system have been successfully integrated with PRSS structure as shown in fig. 10. The structure has been assembled using Al6061 honeycomb structure with the help of end attachments and inserts. Inserts are designed and developed according to ESA standards and end attachments are developed using AU4G. Thrusters have been mounted on each panel with the help of inserts and titanium bolts. Titanium bolts are used for the purpose of high strength and light weight. Four thrusters are mounted on right face of the structure, four on the left side while set of two thrusters are mounted in the middle of each panel for pitch stabilization. Propellant tank is mounted on the inner side of the top panel with the help of inserts and titanium bolts.

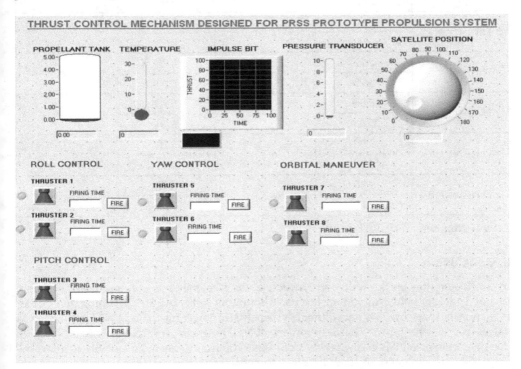

Fig. 11. Thrust Control Panel for PRSS Propulsion System

4.6 Thrust control mechanism

The control panel for the thrust system has been designed on Lab view software as shown in fig. 11 to test the system on ground level. This controller monitors the position of the satellite as well as pressure of the propellant tank and solenoid valves from pressure transducers present in the system. It also observes the impulse bit of the system and the temperature of the propellant tank. Firing time of thrusters is well adjustable on the panel. Ground test of the propulsion system can be monitored by this control system. The test results are listed in table 5.

	Opening Coil Response @ 8bars, 24 VDC (msec)	Minimum Impulse Bit, msec	Opening Coil Response @ 10 bars, 24 VDC, msec	Minumum Impulse Bit, msec
Thruster # 1	2.9	6.20	3.20	6.15
	3.3		2.95	
Thruster # 2	2.95	6.60	3.30	6.15
	3.65		2.85	
Thruster # 3	2.85	6.10	3.50	6.40
	3.25		2.90	
Thruster # 4	2.80	5.80	2.90	6.65
	3.00		2.75	
Thruster # 5	2.77	6.02	3.25	6.25
	3.25		3.00	
Thruster # 6	2.86	6.72	3.10	6.30
	3.86		3.20	
Thruster # 7	2.90	6.56	3.10	6.25
	3.66		3.15	
Thruster # 8	2.80	6.53	3.25	6.40
	3.73		3.15	

Table 5. Minimum Impulse Bit

5. Conclusion

In conclusion, this work results in reduction of the size, mass, power, and cost of system. Use of Titanium bolts, Aluminum Inserts, Aluminum Tank, and PTFE Tubing gives great reduction in mass by 35% and ultimately benefits in lowering the cost. Electric Solenoid valves reduce the power consumption by 40%. The main purpose of this work is to document the potentials of low power Cold Gas Propulsion System adequately to allow the engineers and designers of small satellites to consider it as a practical propulsion system option.

6. Abbreviations

W_i	Initial vehicle weight, Kg
W_f	Final vehicle weight, Kg
W_p	Propellant weight required to produce the given ΔV
ΔV	Velocity increase of vehicle, m/s
g_c	Gravitational constant, 9.8 m/s^2
P	Pressure of the gas, bars
V	Volume of the gas, m^3
m	Mass of the gas, Kg
R	General gas constant, KJ/KgK
T	Temperature of the gas, K
A	Area, m^2
r	Internal radius of the tank, m
t	Thickness of the tank wall, m
σ	Allowable Stresses, MPa
e	Joint Efficiency
σ_h	Hoop Stress
d_x	Length of an element in Cylindrical pressure vessel, m
$\dot{m}$	Mass flow rate of the propellant, Kg/s
V_e	Exit velocity, m/s
$\dot{w}$	Weight flow rate of propellants, N/s
P_e	Exit pressure of the propellant, bars
A_e	Exit Area, mm^2
M_e	Exit Mach number
P_e	Exit pressure, Bars
γ	Specific heat ratio
I_{sp}	Specific Impulse, S
C^*	Characteristics velocity, m/s
P_c	Chamber pressure in the nozzle, Bars
P_t	Pressure at throat, Bars
a_0	Sonic velocity of the gas, m/s
T_c	Chamber temperature, K

7. References

Assad Anis, Design and development of cold gas propoulsion systems for Pakistan Remote Sensing Satellite Systems, 978-1 4244-3300-1, 2008, pg-49-53, IEEE.

Charles D. Brown, Spacecraft propulsion, AIAA series.

DUPONT, SOVA' 134 A, Material Safety Data Sheet, October 2006

European Corporation for Space Standardization (ECSS), ECSS-E-32-02A

Guide book for the design of ASME section VIII Pressure vessels, Third Edition

Handbook of Bolts and Bolted Joints, Edited by John H. Bickford and Sayed Nassar Acknowledgement

Micci, Michael M. and Andrewd, KetsDever, Ed. Micro Propulsion For Small Spacecraft Volume 187, Reston, Virginia: American Institute of Aeronautics and Astronautrics, Inc., 2000.

NASA-STD-5003, Fracture Control Requirement for Payload using the Space Shuttle, 7th October 1996

Wertz, James R. And Wilry J. Larson, Space Mission Analysis and Design, Third Ed. Segundo, California, Microcosm Press, 1999.

Wiley J. Larson, James R. Wertz, Space Mission Analysis and Design, Third Edition, ISBN 1-881883-10-8

Zakirov V., Sweeting M., Erichsen P. and Lawrence T. "Specifics of small satellite propulsion" Part 1, 15th AIAA Conference on Small Satellites, (2001).

Permissions

The contributors of this book come from diverse backgrounds, making this book a truly international effort. This book will bring forth new frontiers with its revolutionizing research information and detailed analysis of the nascent developments around the world.

We would like to thank Boris Escalante-Ramírez, for lending his expertise to make the book truly unique. He has played a crucial role in the development of this book. Without his invaluable contribution this book wouldn't have been possible. He has made vital efforts to compile up to date information on the varied aspects of this subject to make this book a valuable addition to the collection of many professionals and students.

This book was conceptualized with the vision of imparting up-to-date information and advanced data in this field. To ensure the same, a matchless editorial board was set up. Every individual on the board went through rigorous rounds of assessment to prove their worth. After which they invested a large part of their time researching and compiling the most relevant data for our readers. Conferences and sessions were held from time to time between the editorial board and the contributing authors to present the data in the most comprehensible form. The editorial team has worked tirelessly to provide valuable and valid information to help people across the globe.

Every chapter published in this book has been scrutinized by our experts. Their significance has been extensively debated. The topics covered herein carry significant findings which will fuel the growth of the discipline. They may even be implemented as practical applications or may be referred to as a beginning point for another development. Chapters in this book were first published by InTech; hereby published with permission under the Creative Commons Attribution License or equivalent.

The editorial board has been involved in producing this book since its inception. They have spent rigorous hours researching and exploring the diverse topics which have resulted in the successful publishing of this book. They have passed on their knowledge of decades through this book. To expedite this challenging task, the publisher supported the team at every step. A small team of assistant editors was also appointed to further simplify the editing procedure and attain best results for the readers.

Our editorial team has been hand-picked from every corner of the world. Their multi-ethnicity adds dynamic inputs to the discussions which result in innovative outcomes. These outcomes are then further discussed with the researchers and contributors who give their valuable feedback and opinion regarding the same. The feedback is then collaborated with the researches and they are edited in a comprehensive manner to aid the understanding of the subject.

Apart from the editorial board, the designing team has also invested a significant amount of their time in understanding the subject and creating the most relevant covers. They scrutinized every image to scout for the most suitable representation of the subject and create an appropriate cover for the book.

The publishing team has been involved in this book since its early stages. They were actively engaged in every process, be it collecting the data, connecting with the contributors or procuring relevant information. The team has been an ardent support to the editorial, designing and production team. Their endless efforts to recruit the best for this project, has resulted in the accomplishment of this book. They are a veteran in the field of academics and their pool of knowledge is as vast as their experience in printing. Their expertise and guidance has proved useful at every step. Their uncompromising quality standards have made this book an exceptional effort. Their encouragement from time to time has been an inspiration for everyone.

The publisher and the editorial board hope that this book will prove to be a valuable piece of knowledge for researchers, students, practitioners and scholars across the globe.

List of Contributors

Han-Dol Kim, Gm-Sil Kang, Do-Kyung Lee, Kyoung-Wook Jin and Seok-Bae Seo
Korea Aerospace Research Institute (KARI), Republic of Korea

Hyun-Jong Oh
Korea Meteorological Administration (KMA), Republic of Korea

Joo-Hyung Ryu
Korea Ocean Research & Development Institute (KORDI), Republic of Korea

Herve Lambert, Ivan Laine, Philippe Meyer, Pierre Coste and Jean-Louis Duquesne
EADS Astrium, France

Yasser Hassebo
LaGuardia Community College of the City University of New York, USA

Charles R. Bostater, Jr., Gaelle Coppin and Florian Levaux
Marine Environmental Optics Laboratory and Remote Sensing Center, College of Engineering,
Florida Institute of Technology, Melbourne, Florida, USA

Sivakumar Venkataraman
Council for Scientific and Industrial Research, National Laser Centre, Pretoria, South Africa
University of Pretoria, Department of Geography Geoinformatics and Meterology, Pretoria,
South Africa
University of Kwa-Zulu Natal, Department of Physics, Durban, South Africa

Mykhaylo Palamar
Department of Devices and Control-Measurement Systems, Information Technique and
Intelligent Systems Research Laboratory, Ternopil National Technical University, Ukraine

Jianquan Yao, Ran Wang, Haixia Cui and Jingli Wang
Tianjin University, China

Albert Lin
National Space Organization, Taiwan, R.O.C.

Hang Jin, Marc Miska and Edward Chung
Smart Transport Research Centre, Queensland University of Technology, Brisbane, Australia

Maoxun Li
College of Urban Economics and Public Administration, Capital University of Economics
and Business, Beijing, PR China

Yanming Feng
Faculty of Science and Technology, Queensland University of Technology, Australia

Larbi Talbi and Ismail Ben Mabrouk
Université du Québec en Outaouais, Canada

Mourad Nedil
Université du Québec en Abitibi-Témiscamingue, Canada

Assad Anis
NED University of Engineering and Technology, Pakistan

Printed in the USA
CPSIA information can be obtained
at www.ICGtesting.com
JSHW011426221024
72173JS00004B/685

9 781632 392893